Kinetic investigation of different supported catalysts for the poly
under industrially relevant conditions

# Kinetic investigation of different supported catalysts for the polymerization of propylene under industrially relevant conditions

# Dissertation

zur Erlangung des Doktorgrades der Ingenieurwissenschaften
(Dr.-Ing.)

der

Naturwissenschaftlichen Fakultät II
Chemie, Physik und Mathematik

der Martin-Luther-Universität
Halle-Wittenberg

vorgelegt von

Frau Joana Kettner

geb. am 12.09.1985 in Bautzen

Gutachter: Prof. Dr.-Ing. Michael Bartke

Gutachter: Prof. Dr. Markus Busch

Tag der öffentlichen Verteidigung: 14.12.2018

**Bibliografische Information der Deutschen Nationalbibliothek**

Die Deutsche Nationalbibliothek verzeichnet diese Publikation in der
Deutschen Nationalbibliografie; detaillierte bibliographische Daten sind im Internet
über http://dnb.d-nb.de abrufbar.

1. Aufl. - Göttingen: Cuvillier, 2019
  Zugl.: Halle-Wittenberg, Univ., Diss., 2018

© CUVILLIER VERLAG, Göttingen 2019
  Nonnenstieg 8, 37075 Göttingen
  Telefon: 0551-54724-0
  Telefax: 0551-54724-21
  www.cuvillier.de

  ISBN 978-3-7369-7003-8
  eISBN 978-3-7369-6003-9

# Abstract

Scope of the work is the kinetic investigation of two fourth generation Ziegler-Natta catalysts as well as a supported metallocene catalyst for the polymerization of propylene under industrially relevant conditions and the development of simplified phenomenological kinetic models describing the polymerizations. Therein, the influence of different reaction conditions (temperature, pressure, hydrogen concentration) and, in particular, the effect of prepolymerization on catalyst kinetics as well as on polymer characteristics are studied.

The Ziegler-Natta catalysts were investigated under gas-phase conditions in a 5 l horizontal stirred tank reactor operating in semi-batch mode. Both catalysts showed a similar kinetic behavior as well as hydrogen response at the different reaction conditions. Applying a prepolymerization led to an increase in activity at higher reaction temperatures as well as an improved polymer morphology. The impact of prepolymerization is catalyst specific and depends on the catalyst activity reached at main polymerization temperature.

The supported metallocene catalyst was studied under bulk conditions in liquid propylene using a special 250 ml reaction calorimeter. A focus was set on procedure development (in-situ and external prepolymerization) with the target of defined prepolymerization conditions and early access to the kinetic profile. Final kinetic measurements were carried out using the developed external prepolymerization procedure.

Based on the experimental studies, simplified phenomenological kinetic models are developed for each catalyst type enabling the quantitative description of the polymerization reactions at the different reaction conditions including the effect of prepolymerization.

Main hypothesis for the mathematical description of the prepolymerization effect is that particle overheating at the beginning of the polymerization is the major reason for lower activities obtained when no prepolymerization is applied. For particle modeling, a quasi-homogeneous particle model is assumed considering particle growth and particle heat-transfer. As similar kinetic behaviors were observed for both Ziegler-Natta catalysts, the same kinetic model can be used; differences can only be described by the catalyst specific amount of polymerization active component.

A similar kinetic model approach is used to describe the bulk polymerization with metallocene catalyst. Differences regarding the different polymerization regime, in particular monomer concentration in the polymer particle, hydrogen concentration in liquid phase, are considered.

With the estimated sets of kinetic parameters, average activities and average molecular weights can be calculated as well as activity profiles of the catalysts can be quantitatively described at the different reaction conditions.

# Abstract

Im Rahmen dieser Arbeit wird die Kinetik von zwei Ziegler-Natta Katalysatoren der vierten Generation sowie eines geträgerten Metallocen Katalysators für die Polymerisation von Propylen unter industrierelevanten Bedingungen untersucht. Darin wird der Einfluss verschiedener Reaktionsbedingungen (Temperatur, Druck, Wasserstoffkonzentration) und im Besonderen der Einfluss der Prepolymerisation auf den kinetischen Verlauf der Katalysatoren betrachtet sowie Polymereigenschaften untersucht. Des Weiteren werden für jeden Katalysatortyp vereinfachte, phänomenologische kinetische Modelle abgeleitet, mit denen die Polymerisationen mit und ohne Prepolymerisation bei den verschiedenen Reaktionsbedingungen quantitativ beschrieben werden können.

Die Polymerisationen mit Ziegler-Natta Katalysatoren wurden in der Gasphase in einem 5 l horizontal gelagerten Rührkesselreaktor durchgeführt, der in semi-batch Modus betrieben wurde. Beide Ziegler-Natta Katalysatoren zeigten ähnliche kinetische Verläufe sowie ein ähnliches Wasserstoffansprechverhalten. Bei Polymerisationen mit Prepolymerisation wurden deutlich höhere Katalysatoraktivitäten sowie eine verbesserte Polymermorphologie erzielt. Dabei ist der Einfluss der Prepolymerisation katalysatorspezifisch und abhängig von der Katalysatoraktivität.

Die Polymerisation mit Metallocen Katalysator wurde in flüssigen Propylen in einem speziellen 250 ml Reaktionskalorimeter durchgeführt. Es wurden verschiedene Methoden entwickelt (in-situ und externe Prepolymerisation), um definierte Prepolymerisationsbedingungen sowie eine frühe Messung der Kinetik zu ermöglichen. Die kinetischen Untersuchungen wurden mit der entwickelten externen Prepolymerisationsmethode durchgeführt.

Für die mathematische Beschreibung der Polymerisation wird ein quasi-homogenes Partikelmodell angenommen, bei dem Partikelwachstum sowie Partikelwärmebilanz berücksichtig werden. Dabei wird angenommen, dass eine Partikelüberhitzung zu Beginn der Rektion zu einer Verringerung der Katalysatoraktivität führt, wenn die Polymerisation ohne Prepolymerisation durchgeführt wird. Aufgrund der ähnlichen kinetischen Verläufe kann das gleiche Modell für beide Ziegler-Natta Katalysatoren verwendet werden. Unterschiede zwischen den Katalysatoren werden nur anhand der unterschiedlichen Menge an polymerisationsaktiven Zentren wiedergeben. Für die Modellierung der Flüssigphasenpolymerisation mit Metallocen Katalysator wird ein ähnliches kinetisches Modell hergeleitet, bei dem Unterschiede in der Monomer- sowie Wasserstoffkonzentration in der flüssigen Phase berücksichtigt werden.

Mit den ermittelten kinetischen Parametern können gemittelte Katalysatoraktivitäten und Molmassen berechnet sowie die kinetischen Verläufe der Katalysatoren während der Polymerisation bei den verschiedenen Reaktionsbedingungen quantitativ beschrieben werden.

# Danksagung

Mein besonderer Dank gilt an erster Stelle meinem Doktorvater Prof. Dr.-Ing. Michael Bartke. Ich möchte mich für das interessante, industrieorientierte Thema und für die ausgezeichnete Betreuung während der Zeit an der Martin-Luther-Universität Halle-Wittenberg sowie danach bis zur Abgabe meiner Dissertation bedanken. Besonders geschätzt habe ich die Möglichkeit der selbständigen, abwechslungsreichen und spannenden Arbeit im Polymerisationslabor sowie die vielen Gelegenheiten, mein Wissen bei Weiterbildungen und auf Fachkonferenzen zu erweitern.

Ich bedanke mich recht herzlich bei Prof. Dr. Markus Busch für sein Interesse an meiner Arbeit und die Übernahme des zweiten Gutachtens.

Ich möchte der Lummus Novolen Technology GmbH für die finanzielle Unterstützung sowie für die sehr angenehme Zusammenarbeit danken. Insbesondere ein herzliches Dankeschön an Dr. Thorsten Sell, Yvonne Denkwitz, Dr. Martin Dietrich sowie Dr. Oliver Ruhl, für die fachliche Beratung und Unterstützung sowie die Durchführung der vielen Polymeranalysen.

Vielen Dank an meine Kollegen der Arbeitsgruppe Polymerreaktionstechnik für die gegenseitige Unterstützung. Mein besonderer Dank gilt Dr. Thomas Kröner, der mir während der gemeinsamen Doktorandenzeit alles Wichtige zur Polymerisation im Labor beigebracht hat und auch danach mit seinem Rat zur Seite stand.

Prof. Dr.-Ing. habil. Dr. h.c. Ulrich und Prof. Dr.-Ing. habil. Hahn danke ich für die Möglichkeit in ihren Arbeitsgruppen die Analysen der Polymerproben durchführen zu können. Besonders bedanke ich mich herzlich bei Jenny Bienias für die vielen Dichte- und Porositätenmessungen sowie bei Herrn Lebek für die Erklärung und Durchführung der Rasterelektronenmikroskopie.

Darüber hinaus möchte ich mich bei Herrn Dr. Lother Karrer von der BASF SE für die Beratung bei der Anpassung der Monomerreinigung in unserem Labor bedanken.

Ich möchte mich ebenfalls bei meinen jetzigen Arbeitskollegen bei Borealis, insbesondere bei Dr. Vasileios Kanellopoulos, für ihre Unterstützung und ihr Verständnis während der Phase des Zusammenschreibens bedanken.

Der größte Dank gilt meinen Eltern, meinen Großeltern und meinem Bruder. Ich bedanke mich von ganzem Herzen für euren Rückhalt und für die unentwegte Unterstützung.

# Table of content

# Nomenclature

| Abbreviations | Description | Unit |
|---|---|---|
| Avg | Average | |
| CCD | Chemical composition distribution | |
| DP | Degree of prepolymerization | |
| DSC | Differential scanning calorimetry | |
| E/P-RACO | Ethylene/propylene random copolymer | |
| FBR | Fluidized bed reactors | |
| GP | Gas-phase | |
| HF | Heat flow | W |
| HP | Homopolymer | |
| lq | Liquid | |
| MC | Metallocene | |
| MFC | Mass flow controller | |
| MFR | Melt mass flow rate | g/10 min |
| MW | Molecular weight | |
| MWD | Molecular weight distribution | |
| PDI | Polydispersity index | |
| PE | Polyethylene | |
| PP | Polypropylene | |
| PP-HECO | Polypropylene heterophasic copolymer | |
| Prop or C3 | Propylene | |
| PS | Particle size | |
| PSD | Particle size distribution | |
| RMP | Raw material purification | |
| SEM | Scanning electron microscopy | |
| Silane | Cyclohexyl-dimethoxy-methylsilane | |
| TEA | Triethylaluminum | |
| TIBA | Triisobutylaluminum | |
| VLE | Vapor-liquid equilibrium | |
| ZN | Ziegler-Natta | |

| Greek letters | Description | Unit |
|---|---|---|
| $\alpha$ | Heat transition coefficient | $W/(m^2 \cdot K)$ |
| $\beta$ | Lattice constant | - |
| $\delta$ | Solubility parameters of the monomer and polymer, resp. | $(cal/cm^3)^{0.5}$ |
| $\eta$ | Dynamic viscosity | $Pa \cdot s$ |
| $\vartheta$ | Temperature | $^\circ C$ |
| $\lambda$ | Heat conductivity coefficient | $W/(m \cdot K)$ |
| $\rho$ | Density | g/l |
| $\upsilon_m$ | Molar volume of the liquid monomer | l/mol |
| $\phi$ | Volume fraction of the permeant | - |
| $\chi$ | Flory-Huggins interaction parameter | - |

| Capital letters | Description | Unit |
|---|---|---|
| $\bar{A}$ | Average activity | kg/(g·h) |
| $\dot{Q}$ | Heat flow | W |
| A | Activity | kg/(g·h) |
| A | Area | m² |
| C | Catalyst | - |
| D | Dead polymer chains | - |
| $E_A$ | Activation energy | J/mol |
| K | Crystallinity | % |
| M | Monomer | - |
| Mn | Number average molecular weight | g/mol |
| Mw | Weight average molecular weight | g/mol |
| N | Number of polymer particles | - |
| Nu | Nusselt number | - |
| P | Growing polymer chains | - |
| Pr | Prandtl number | - |
| R | Universal gas constant | J/(mol·K) |
| Re | Reynolds number | - |
| $R_M$ | Overall reaction rate of monomer consumption | mol/(l·s) |
| $R_p$ | Overall reaction rate of polymerization | mol/(l·s) |
| S | Species | - |
| T | Temperature | K |
| V | Volume | l |
| $V_R$ | Reaction volume | l |
| Y | Dormant polymer chains | - |
| $\Delta_R H$ | Reaction enthalpy | J/mol |
| $\Delta H_m$ | Specific melt enthalpy | J/mol |

| Small letters | Description | Unit |
|---|---|---|
| c | Concentration | mol/l |
| $c_M$ | Monomer concentration | mol/l |
| $c_M{}^*$ | Equilibrium monomer concentration in polymer | mol/l |
| $c_p$ | Specific heat capacity | J/(g·K) |
| d | Diameter or distance | m |
| g | Growing factor | - |
| k | Heat transfer coefficient | W/(m²·K) |
| $k^*$ | Henry constant | mol/(atm·l$_{amorph}$) |
| $k^0$ | Pre-exponential factor | l/(mol·s) or 1/s |
| $k_x$ | Rate constant of reaction x (x=act, p, tr, dorm, des, des,temp) | l/(mol·s) or 1/s |
| m | Mass | g |
| $\dot{m}$ | Mass flow | kg/h |
| n | Moles | mol |
| p | Pressure or partial pressure | bar |
| $p^0$ | Saturation vapor pressure | bar |
| $r_x$ | Reaction rate of reaction x (x=act, p, tr, dorm, des, des,temp) | mol/(l·s) |
| t | Time | h |
| u | Relative velocity | m/s |
| w | Weight fraction | wt% |
| $x_{active}$ | Semi-empirical factor describing active fraction of Ti for polymerization | - |

| Indices | Description | Indices | Description |
|---|---|---|---|
| * | Active site or equilibrium | i | Particle size fraction |
| 0 | Initial conditions at reaction start | J | Jacket |
| act | Activation | L | Liquid phase |
| bulk | Bulk phase | M | Monomer |
| c or crit | Critical | n | Chain length |
| Cat | Catalyst | P | Polymer particle |
| des | Deactivation | p | Propagation |
| des, temp | Thermal deactivation | Pol | Polymer |
| dev | Deviation | PP | Polypropylene |
| dorm | Dormant | R | Reactor |
| GP | Gas-phase | tr | Transfer |
| HF | Heat flow | | |

| Chemical formula | Description | Chemical formula | Description |
|---|---|---|---|
| $C_3H_6$ | Propene / propylene | $N_2$ | Nitrogen |
| CO | Carbon monoxide | $O_2$ | Oxygen |
| $CO_2$ | Carbon dioxide | $SiO_2$ | Silicon dioxide |
| $H_2$ | Hydrogen | Ti | Titanium |
| $H_2O$ | Water | $TiCl_4$ | Titanium tetrachloride |
| $MgCl_2$ | Magnesium dichloride | Zr | Zirconium |

# 1 Introduction

## 1.1 Polypropylene – markets and applications

Polypropylene (PP) belongs to the group of polyolefins and is next to polyethylene (PE) the second most important polymer for daily life applications. In Europe, the total polymer demand in 2014 reached 47.8 million tons, wherein polypropylene had the largest marked share of 19.2 % (9.18 million tons) compared to PE-LD and PE-LLD with 17.2 % and PE-HD with 12.1 %, see Figure 1 [1]. The global demand of polypropylene in 2013 was 55.1 million tons [2]. More than the half of the polypropylene is used as packaging material such as flexible packaging made from polypropylene films or rigid packaging e.g. containers or caps. Further applications are polypropylene fibers and consumer goods, which reached a marked share of 12 % in each case. Other applications are in the automobile industry, in the building and construction sector as well as in the electro and electronic sector. Also the application of special materials like fiber-reinforced or foamed polypropylene and copolymers (e.g. with ethylene) as technical thermoplastics is increasing [2],[3]. The market research institute Ceresana expected until 2021 an annually revenue growth for polypropylene of 5.8 %. Especially due to the opening of new application areas, the substitution of other materials and the development of bio-based polypropylene materials will lead to the high growth potential for the polypropylene market [2].

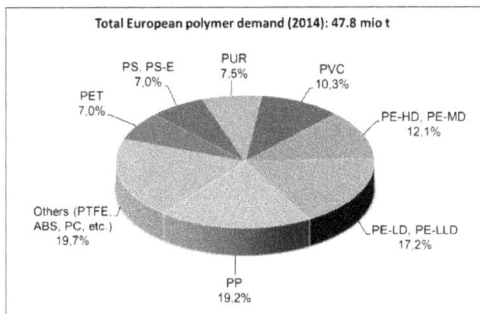

Figure 1: European polymer demand by type 2014 [1]

In 2012, 65 million tons of polypropylene were produced worldwide. Half of the production capabilities are located in Asia, 17 % in Europe, 14 % in North America and 12 % in Middle East [4]. The major polypropylene producers are LyondellBasell, Sinopec, PetroChina, Braskem, Borealis, Sabic, Exxon Mobil, Reliance Industries, Total Petrochemicals and Formosa Plastics Corporation [5].

## 1.2 A brief description of polypropylene microstructure and properties

The basic repeating unit propylene consists of three carbon atoms, where the methyl group is chiral leading to different stereochemical configurations of the methyl group in the polymer chain. The three main configurations of polypropylene are [6]:

a) *Isotactic polypropylene:*

 The methyl groups are oriented on the same side of the backbone of the polymer chain.

b) *Syndiotactic polypropylene:*

 The methyl groups alternate regularly from side to side in the polymer chain.

c) *Atactic polypropylene:*

 The methyl groups are randomly oriented in the polymer chain.

The different stereoisomers of PP are shown schematically in Figure 2. The most common form in industry is isotactic polypropylene, which is easily produced with state-of-the-art heterogeneous Ziegler-Natta and metallocene catalysts [5],[6],[7]. It is characterized by a high crystallinity (60-70%) and a high melting temperature in the range of 160-165°C. Syndiotactic polypropylene is also semi-crystalline but is less stiff than isotactic PP and has a lower melting temperature of 130 °C. Atactic PP is amorphous and a rubbery material. Generally, it is an undesired product but some applications can be found in the electro industry as casting compound or as insulating or sealing material [3],[8].

Modern coordination catalysts used in industry are able to produce very high isotactic PP with only minor atactic fractions (the catalysts are described in more detail in chapter 2.1). The polymer chains have a high regioregularity, wherein a 1,2 insertion of the monomer and head-to-tail enchainment is favored. Regioerrors caused by 2,1 insertion of propylene create irregularities along the polymer chain leading to a decrease in crystallinity and a lower melting temperature [6],[7],[9].

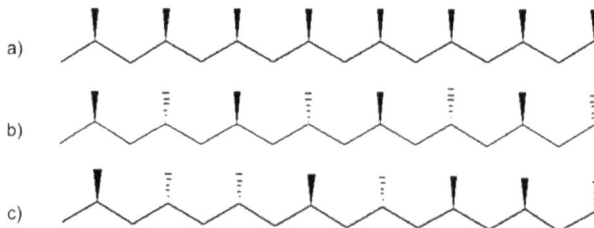

Figure 2: Main polypropylene types: a) isotactic, b) syndiotactic, c) atactic [7]

The properties of PP are also dependent on the molecular weight and the molecular weight distribution (MWD). They are influenced by the process conditions (e.g. production of unimodal

2

or bimodal MWD), the used catalyst type and additives such as hydrogen which acts as chain transfer agent [5]-[7],[10]. Bimodal polymers can be produced in different ways [11]: In multi-stage processes using same or different reactor types with different hydrogen concentrations in each stage (e.g. Borstar PP process), in multi-zone reactors (e.g. Spherizone PP process, both see chapter 1.3) or with multisite catalysts [12]. The control of the MWD allows the production of polymers with defined properties as well as new property combinations (Figure 3). With Ziegler-Natta catalysts produced polymers have a relatively broad MWD with a polydispersity index (PDI) in the range from 4 to 20 [5],[13], whereas polymers produced with metallocene catalysts have a narrow MWD with a PDI of approx. 2 [7],[14] (see chapter 2.1).

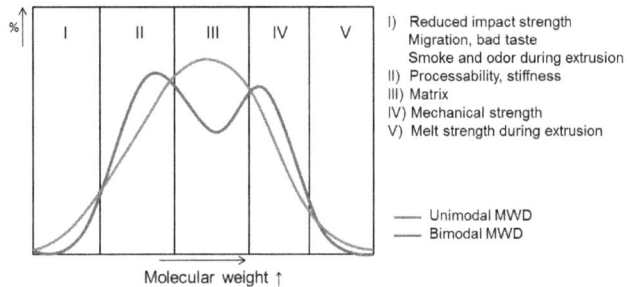

Figure 3: Molecular weight distributions of unimodal and bimodal polymers (PE and PP) and its impact on polymer properties [7], [11]

Commercial PP grades are grouped in three different grade classes:

- Homopolymer
- Random copolymer
- Heterophasic or impact copolymer

Polypropylene homopolymers are characterized by a good tensile strength and stiffness but a poor (low-temperature) impact resistance and film clarity [3],[5],[6].

Random copolymers are copolymers of propylene and ethylene (E/P-RACO). The ethylene content is in the range of 1 to 8 wt%. The ethylene molecules are randomly distributed in the polypropylene backbone. Ethylene is the most commonly used comonomer to modify the properties of propylene but also 1-butene or terpolymers of ethylene and 1-butene are used. In E/P-RACO, optical properties (good film clarity), impact resistance and flexibility are improved. In contrast, with increasing comonomer content, crystallinity, stiffness and melting temperature decrease [3],[6],[11].

Heterophasic copolymers (PP-HECO) consist of a continuous phase (matrix) of homo-polypropylene or random copolymer and an elastomeric phase of amorphous, rubber-like propylene/ethylene copolymer. The ethylene content in the elastomer is typically in the range of 20 to 60 wt%. The heterophasic copolymer consists of 10 to 20 wt% of elastomer which is not

3

miscible with the matrix phase [11]. Main advantages of the heterophasic copolymer are the improved low-temperature impact strength, a high thermal stability and an improved blush resistance [6]. Beside these three basic types, additional modification of the polypropylene properties can be achieved by mixing with other polymers, with filling or reinforcing materials, pigments or by the chemical reaction with the final product [3].

Polypropylene is characterized by good electrical insulation properties and is highly resistant to a wide range of chemicals, except of strong acids and oxidants. The density of PP is with approx. 0.85 g/cm³ (atactic PP) to 0.94 g/cm³ (isotactic PP) low, which makes it to an important material for packaging applications or materials where low weights are desired. As polypropylene is a thermoplastic, it can be melted and shaped into a desired form and subsequently remelted and reshaped into other forms. In terms of sustainability it is therefore a good recycling material. Due to its good mechanical and thermal properties, it can be processed in a variety of ways, mainly by injection molding, fiber extrusion and film extrusion [3],[8].

## 1.3   Processes for the industrial production of polypropylene

Industrial polymerization processes for the production of polypropylene can be categorized into following main groups:

- Gas-phase polymerization
- Bulk polymerization in liquid monomer
- Slurry polymerization in inert diluent

PP is produced either in one phase or in combination of different phases (hybrid or mixed-phase processes). The development of PP polymerization processes is strongly related to advances in catalyst development (see chapter 2.1) and the demand of improved product performance. In early polymerization processes, additional process steps for the removal of catalyst residuals and atactic PP were necessary [6]. The development of more active and highly stereoselective catalysts led to the design of more efficient processes where such treatment steps are not required anymore [7],[9],[15]. Modern polyolefin processes are very efficient and produce large quantities of polymer. Starting with 5 kt/y in 1963, world scale plants have grown from 80 kt/y in 1980 to more than 750 kt/y for the newest plants [7].

Liquid phase processes use autoclaves or (more common today) loop reactors. Gas-phase polymerizations are carried out in fluidized bed reactors (FBR) or in stirred bed reactors. In FBR, a gaseous stream of monomer and nitrogen is used to fluidize the polymer particles in the reactor, whereas in stirred gas-phase reactors mechanical stirring is used to suspend the polymer particles. The stirred gas-phase reactors can further divided into horizontal and vertical reactor configurations. The polymerization can be carried out in one reactor, but typically two or more reactors of same or different configuration are connected in series in order to produce polymer products with a more complex structure. Propylene homopolymer and random copolymers are produced in various configurations of gas-phase and bulk reactor systems. Heterophasic copolymers are only produced in an additional gas-phase reactor (often FBR) because of the stickiness of the polymer and the solubility of the propylene/ethylene rubber phase in the monomer and diluent [7].

A detailed description of the several reactor configurations as well as polyolefin processes can be found in [6], [7] and [15]. Table 1 shows an overview of the current commercial PP polymerization processes with the reactor configurations. The named technology suppliers are the current technology licensers. In the following, an overview of the main polymerization processes for the production of PP will be given.

Table 1: Industrial polypropylene processes, based on [15]

| Process | Technology supplier | Reactor type | | Mode of operation |
|---|---|---|---|---|
| | | Homopolymer | Impact copolymer | |
| Unipol PP | Grace | Fluidized bed reactor (FBR) | FBR, gas | Condensed gas phase/gas |
| Novolen | Novolen Technology Holdings | Vertical stirred bed gas-phase reactor | Vertical stirred bed gas-phase reactor | Gas/gas (non-condensed) |
| Innovene PP | Ineos | Horizontal stirred gas-phase reactor | Horizontal stirred gas-phase reactor | Gas/gas |
| Horizone | JPP (Japan Polypropylene Corp.) | Horizontal stirred gas-phase reactor | Horizontal stirred gas-phase reactor | Gas/gas |
| Spherizone | LyondellBasell | Multizone circulating reactor | FBR, gas | Gas/gas |
| Spheripol | LyondellBasell | Loop reactor | FBR, gas | Bulk/gas |
| Borstar PP | Borealis | Loop reactor and FBR | FBR, gas | Bulk+gas/gas |
| Hypol I | Mitsui | Stirred autoclave reactors | FBR, gas | Bulk/gas |
| Hypol II | | Loop reactors | FBR, gas | Bulk/gas |
| Exxon Mobile | Exxon Mobile | Loop | | Bulk |
| Slurry | Several | Series of autoclaves | Series of autoclaves | Slurry |

Slurry (inert diluent) processes

The first commercial processes for the production of PP were slurry processes, where the polymerization took place in an aliphatic hydrocarbon such as hexane as inert diluent. Many different slurry processes were developed in the early 1970's. The Ziegler-Natta catalysts available at this time (1st and 2nd generation ZN catalysts, see section 2.1.1) had low activity and produced polymer with large amounts of atactic polymer. Therefore, processes include a series of CSTR's of up to 5 to 7 reactors in order to achieve good monomer conversion. Additional post-reactor treatment was necessary to remove catalyst residuals (deashing) and atactic polymer. Diluents used in the different variations of slurry processes ranges from C6 to C12 hydrocarbons [6],[7],[15]. An overview of early PP manufacturing processes is given by Moore [6].

Diluent slurry processes are very cost intensive because of the high number of equipment (workup section for diluent) and the costly operation of the plant. Over the past few decades, slurry processes have been replaced by more efficient bulk and gas-phase processes. However, there are still remaining diluent slurry plants in operation producing specialty polymers, e.g. high-crystallinity polypropylene (HCPP) [15].

## Bulk (liquid propylene) processes

Polymerization processes carried out in liquid propylene are known as "bulk" or "liquid pool" processes. Advantages to the diluent slurry process are that the usage of liquid monomer as reaction medium increases the polymerization rate due to the higher monomer concentration. Propylene can easily separate from the polymer by flushing and no extensive diluent recovery system is needed. Polymerization processes are performed in continuous stirred tank reactors or loop reactors. They are used to produce homopolymer and random copolymers (<8 wt% ethylene). The first bulk process was developed by Dart Industries in the 1960's, also called Rexene, Rexall or El Paso bulk process. Other bulk processes are Exxon Sumitomo and early process from Mitsui. All of them uses CSTR's or stirred autoclaves [6]. Chevron Phillips Chemical developed the "loop slurry" process for the production of HD-PE, where polymerization is carried out in a series of pipe reactors (up to 8 legs) [5],[6]. The loop design provides a maximum surface area improving the heat removal from the polymerization reaction leading to an increase of the reactor throughput. They were also adapted very quickly to PP processes. Nowadays, loop reactors are used to produce about 50% of all commercial polyolefins [9]. Therein, single bulk configurations are often combined with further gas-phase reactors in order to provide a full range of polymer grades (see section hybrid processes).

## Gas-phase processes

Gas-phase processes are in comparison to slurry processes (diluent or bulk) economically and energy efficient. The polymer can easily separate from the monomer since monomer is in gas-phase. There are no diluents which have to be separated and treated or to flash off large amounts of monomer like in bulk phase. This leads to a simple plant configuration with lower investment and operational costs. One drawback of gas-phase processes is the limited heat transfer capacity in comparison to slurry processes. In order to remove the heat from the system, gas is removed from the reactor, condensed in an external heat exchanger and recycled into the reactor (condensation cooling). When liquid monomer is injected into the reactor, it immediately evaporates leading to a further cooling effect. Polypropylene is produced in gas-phase processes either in fluidized bed reactors or in continuous stirred bed reactors of different configurations [6],[7],[15].

The Unipol polypropylene process, originally developed for the gas-phase polymerization of ethylene by Union Carbide (later part of Dow Chemical, now licensed by Grace), utilizes

6

fluidized bed reactors for the production of polypropylene. Currently 48 operating lines and 15 reactor lines are using the Unipol process technology accounting for 17 % of the global propylene output [16]. A scheme of the process is shown in Figure 4. A single large fluidized bed reactor is used for the production of homopolymer and random copolymer. A second smaller FBR can be connected in series for the production of heterophasic (impact) copolymer.

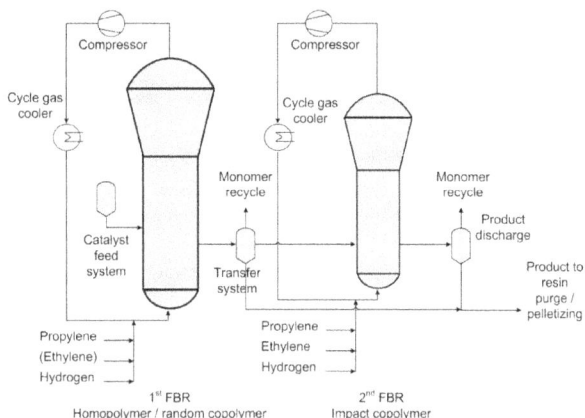

Figure 4: Scheme Unipol PP process [17]

Fresh, purified monomer as well as recycled monomer are fed to the reactor bottom. Therein, the monomer is fed in condensed mode, where 10 to 12 % of propylene is liquid propylene. Catalyst, cocatalyst and donor are fed via a feeding system into the reactor. The polymerization takes place in the reaction zone, where the gas stream is circulated through the bed. The upper section of the reactor is wider in order to reduce the gas velocity and particle entrainment. The gas stream is removed at the top of the reactor, goes through a compressor and is than cooled down in an external heat exchanger to remove the reaction heat. The first reactor operates at 60°C to 70°C and in a pressure range of 25 to 30 bar. The residence time is typically one hour. Monomer is periodically removed from the reactor and separated in a series of high-pressure gas/solid separators. The separated gas is recycled back and the polymer goes through further processing or is transferred to the second FBR. The second reactor is smaller because only 20 % of the reaction takes place. Also copolymerization is carried out at lower temperatures and pressures. The reactor as well as the recycle cooler system is operated only in gas-phase [7],[9].

The Novolen process uses one or two identical vertical stirred bed gas-phase reactors. The process was originally developed by BASF in 1967 and is now licensed by Lummus Novolen Technology holding CB&I [18]. The reactors can be operated very flexible in parallel or cascade mode. The production of homopolymer and random copolymer can either carried out in a single reactor, parallel in two reactors or in two reactors in cascade mode depending on required

7

capacity and product range. For the production of heterophasic copolymer, the two reactors are operated in the cascade mode, wherein the homopolymer (or random copolymer) matrix is produced in the first reactor and in the ethylene-propylene rubber phase is produced in the second reactor. Reactor cooling is achieved by flash evaporation of liquefied reactor gas (condensed in an external heat exchanger) which is mixed with fresh feed and injected into the reactor. The polymer/gas mixture is discharged from the reactor through dip tubes and transferred to the other reactor or to the powder and gas separation system. Unreacted monomer is compressed and recycled into the reactors. The PP powder is transferred via gravity flow to the purge silo, where residual propylene is removed by flushing with nitrogen. Finally, the polymer powder is fed into the extruder, where it is converted together with additives into pellets [18],[19]. A scheme of the Novolen process is shown in Figure 5.

The Innovene PP process (Ineos) and the Horizone process (Japan Polypropylene Corporation) are based on horizontal stirred bed reactors. The technology was developed by Amoco/Chiso in the 1970's. In the Horizone process two reactors are arranged one above the other, wherein the polymer powder flows under the influence of gravity into the second reactor. In the Innovene process, the two reactors are on the same level requiring a powder transfer system. Both processes can be run in cascade or parallel modes. The cascade mode is used to produce impact copolymer. Heat removal is accomplished by evaporating of liquid propylene injected over the powder bed. The advantage of the reactor configuration is its plug-flow characteristic with a narrow powder residence time distribution [7].

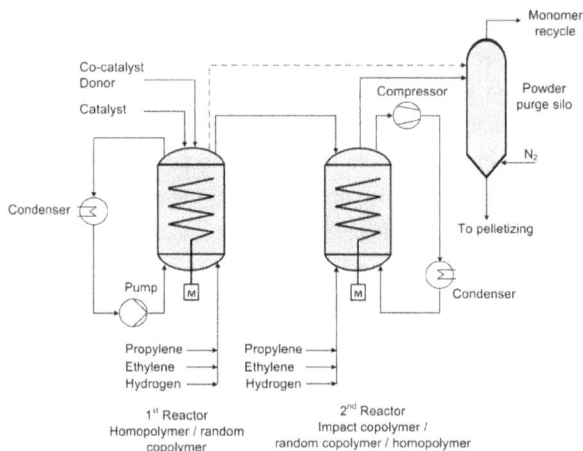

**Figure 5: Scheme Novolen® gas-phase process [19]**

## Hybrid processes

Hybrid or mixed-phase processes combine bulk and gas-phase processes for the production of the full range of polypropylene products (HP, RACO and HECO). The process is divided into one block making homopolymer and random copolymer and a second block producing sticky heterophasic copolymers. While the first block often uses loop reactors or combination of loop reactors and FBR, the second block only consists of a FBR for the reasons described above. The two blocks are separated by a kind of gas lock/ flash system in order to achieve a better control over the polymer properties of the different phases. Advantage of the hybrid processes is the production of a wide range of polymer grades but high capital investment is required [7],[9].

The dominant process is the Spheripol process by LyondellBasell [20], where roughly one third of the world's PP production is based on [9]. Figure 6 shows a scheme of the process. It consists of one to two loop reactors filled with liquid propylene for the production of homopolymer and random copolymer. A small loop reactor is used to prepolymerize the catalyst before it enters into the main loop reactor. Typically, two loop reactors are used in order to narrow the residence time distribution making more uniform polymer product. For the production of impact copolymers a fluidized bed reactor is connected in series. Between the liquid and gas-phase section, a high temperature flush separation is installed in order to depressurize and flush the propylene/PP slurry and to remove $H_2$ for a better control of the molecular weight distribution in the copolymer reactor. Unreacted monomer is recovered from the system and is completely recycled back into the reactors [7].

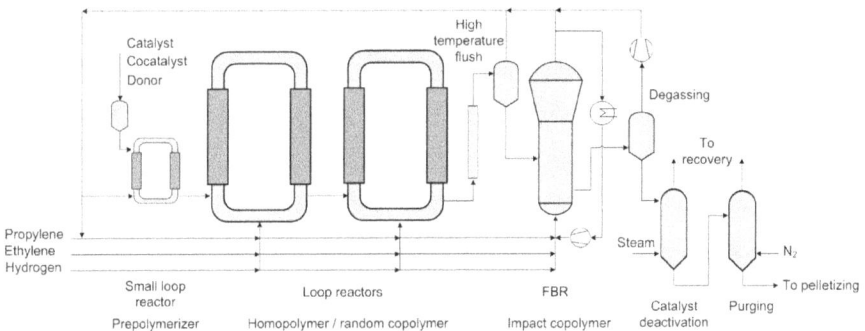

Figure 6: Scheme Spheripol process [7] and [9]

In the Spherizone process, launched by LyondellBasell in 2004, a unique multizone circulating reactor is used instead of the loop reactors which operates in gas-phase [21]. The reactor consists of two interconnected reaction zones separated by a cyclone at the top of the reactor. The riser (first reaction zone) is operated with high gas velocity and acts like a fast fluidized bed. In the cyclone, the polymer particles are separated from the gas and enter the second reaction zone (downer) which can be characterized as a moving packed bed. The two zones can be operated at different conditions ($H_2$ and co-monomer concentrations) where polymer with bimodal properties can be achieved [7],[9],[21].

The Mitsui Hypol process consists of two stirred autoclave reactors working under liquid pool conditions followed by two fluidized bed gas-phase reactors. The gas-phase reactors are stirred fluidized bed reactors with wall scrapers in order to eliminate fouling and sheeting problems. They claimed to achieve higher rubber contents in the production of impact copolymers compared to conventional FBR. The two gas-phase reactors can also be used for the production of homopolymers with a wider range of polymer properties. In the Hypol II process, the stirred autoclaves are replaced by loop reactors enabling a higher throughput and a reduction of capital costs [7],[22].

A third major player for slurry/gas-phase processes is Borealis. They provide the Borstar process technology for PE and PP [23]. The basic design of the Borstar PP process is shown in Figure 7. The catalyst is firstly prepolymerized in a small prepolymerization loop reactor. Homopolymer and random copolymer are produced in a loop reactor followed by a fluidized bed reactor. With this configuration a bimodal polymer matrix can be achieved due to the different reaction conditions. For high impact copolymers one (or two) additional FBR can be connected in series. This FBR are smaller compared to the FBR in the first stage because of the higher reaction rates and typically shorter residence times [9],[23].

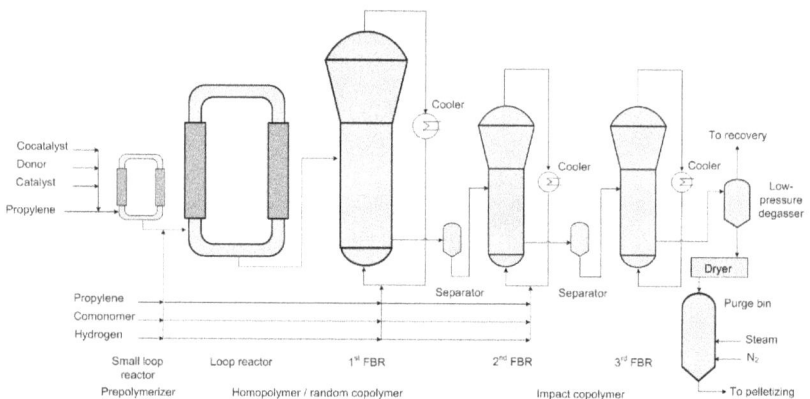

Figure 7: Scheme Borstar PP process (four-reactor setup) [9]

# 2 Literature and scientific background

## 2.1 Catalysts for the polymerization of propylene

All polyolefins, except of low density polyethylene (radical polymerization at high pressure), are produced by coordinative polymerization with organometallic catalysts. There are three different main types of coordination catalysts which are used industrially for the polymerization of polyolefins: Ziegler-Natta catalysts (ZN catalysts), Phillips catalysts and Metallocene catalysts (MC catalysts) [5]-[7],[15]. A fourth type are late-transition metal catalysts, discovered by Brookhart and researchers from DuPont in the early 1990s, which have not found commercial applications yet. They will be therefore not further discussed and it is referred to [7].

The catalysts can be further classified into homogeneous and heterogeneous catalysts. Homogeneous catalysts are soluble in the reaction medium. Examples are homogeneous ZN catalysts based on vanadium compounds and MC catalysts. Heterogeneous catalysts are insoluble in the reaction medium. All supported ZN catalysts, Phillips catalysts as well as supported MC catalysts belong to this group. Main reasons for application of heterogeneous catalysts are the lower viscosity of polymer dispersions in comparison to polymer solutions of same polymer content (better mixing and improved heat removal) as well as a good morphology control [7],[11]. Support materials often used are magnesium dichloride ($MgCl_2$), Silica ($SiO_2$), in some cases also polymeric supports or self-supporting systems are reported [6],[7],[24],[25].

The majority of polypropylene (approx. >97 %) as well as huge quantities of polyethylene (approx. 43 %) are commercially produced with supported Ziegler-Natta catalysts. Phillips catalysts are enormously important for the production of polyethylene (with approx. 20-25 % the global polyethylene production) [5],[26]. The use of MC catalysts for the production of polyolefins increases rapidly due to their high activity, their excellent control of microstructural properties and the ability to implement them into existing production plants. But due to higher costs, they are commercially still less important in comparison to the other catalyst systems. Nevertheless, it is expected that their importance will increase in the next decades [7],[26].

## 2.1.1 Ziegler-Natta catalysts

In general, Ziegler-Natta catalysts are composed of two components: a transition metal halide of transition metals from group IV to VIII of the periodic table of elements such as $TiCl_3$ and $TiCl_4$ and an organometallic compound (metal alkyl) with a base metal from group I to III. The latter acts as cocatalyst or activator and is required to form the active species in the catalyst. Typical cocatalysts are alkyl aluminum compounds such as trimethylaluminum (TMA), triethylaluminum (TEA) or diethylaluminum chloride (DEAC) [5]-[7].

Homogeneous ZN catalysts are based on vanadium compounds and are used for the production of ethylene-propylene-diene elastomers (EPDM) [7],[15].

Polymer produced with heterogeneous (supported) ZN catalysts typically show broad molecular weight distributions. It is commonly believed that this is due to different kinds of active sites in the catalyst system. Each site is able to produce polymer populations with different average microstructural properties - different number average and weight average molecular weights ($M_n$ and $M_w$), average comonomer content and different regio- and stereoregularity. The resulting polymer product is characterized by a broad MWD with polydispersity indices in the range from 4 to 20 (or more common 6-15 [5],[13]) and a broad chemical composition distribution [5],[7],[13],[27],[28].

### History and development of heterogeneous Ziegler-Natta catalysts

ZN catalysts were discovered in 1953 by Karl Ziegler, who polymerized high molecular weight linear polyethylene by a combination of titanium tetrachloride ($TiCl_4$) and trialkyl aluminum ($AlR_3$) [29]. One year later, Giulio Natta produced isotactic polypropylene (with low isotacticity of 30 % to 40 %) using the same type of catalyst [30a,b]. He found out soon that the usage of crystalline $TiCl_3$ instead of soluble $TiCl_4$ (with $AlR_3$ or $AlEt_2Cl$) lead to a highly isotactic PP with 80 % to 90 % isotactic content [30c].

Since the development of the first ZN catalyst, continuous catalyst improvements were done (and are still going on) concerning to higher activities and higher stereoselectivities as well as better control of polymer morphology. ZN catalysts are generally classified in terms of generations according to their chronological order of development. A summary of the different ZN catalysts generations with their main characteristics is given in Table 2. A detailed description of the catalyst development is given e.g. by Malpass and Band [5] and Moore [6].

### First generation catalysts

Based on the first developed catalyst $TiCl_3/AlEt_2Cl$ from Natta [30c] it was discovered that a prolonged ball milling of aluminum reduced $TiCl_3$ or mixtures of $TiCl_3$ and $AlCl_3$ lead to much higher activities than pure $TiCl_3$ [31]. However, the low productivity and stereoselectivity, a poor polymer morphology as well as the additional steps to remove catalyst residuals and separate atactic polymer were the main drawbacks [6].

### Second generation catalysts

In the early 70's, efforts were placed to increase the accessibility to the Ti atoms. This was done e.g. by Solvay by increasing the surface of $TiCl_3$ catalyst [32],[33]. The final catalyst showed a fivefold productivity and isotacticity of around 95 %. The removal of catalysts residues by deashing was still necessary.

Third generation catalysts

As 3rd generation Ziegler-Natta catalysts, catalysts with high surface supports bearing surface functional groups are described. The functional groups improve binding of the transition metal compounds. It was found that catalysts based on activated $MgCl_2$ are very active for PE and PP [34]. The usage of the catalyst was firstly limited to the PE production because only low isotacticity was achieved for PP. This problem was overcome by comilling $MgCl_2$ and $TiCl_4$ with a Lewis base as internal donor and combining this with an aluminum trialkyl ($AlR_3$) as cocatalyst and a second Lewis base as external donor during the polymerization [35]. Internal and external donors are acting as stereoselective control agents leading to an increase of the isotacticity of the polymer. The catalyst could now remain in the product but, with the used donors, atactic polymer had still to be removed [6].

Fourth generation catalysts

4th generation catalysts are $TiCl_4/MgCl_2$ supported catalysts with new combinations of internal and external electron donors as alkylphthalates (diester) and alkoxysilanes [36]. TEA is used as cocatalyst. They were developed in the beginning of the 80's and are highly active and stereoselective catalysts providing isotacticities up to 98 - 99 % as well as spherical polymer morphologies. The catalyst remains now in the polymer and no further step to remove atactic polymer is needed [5],[6].

Fifth generation catalysts

In the second half of 80's, a new type of electron donor (1,3-diethers) was introduced. The new electron donors are used as internal components, which provide very high activities and isotacticities without using any external donor [37]. A second family of the 5th generation catalysts (sometimes also referred to the 6th generation) uses succinates as internal donor [13].

Table 2: Development of heterogeneous Ziegler-Natta catalysts, based on [6] and [5]

| Generation | Catalyst composition | Productivity $kg_{PP}/(g_{cat})$ | Isotacticity wt% | Remarks |
|---|---|---|---|---|
| 1st | $TiCl_3*AlCl_3$+DEAC | 0.8-1.2* | 90-94 | No morphological control (polymer powder and flakes), deashing and atactic removal |
| 2nd | $TiCl_3$+DEAC | 10-15** | 94-97 | Morphology control possible (spherical polymer particles), deashing |
| 3rd | $MgCl_2/TiCl_4$/Ester+$AlR_3$/Ester | 15-30** | 90-95 | Very low stereoselectivity, atactic removal, no morphology control (polymer powder and flakes) |
| 4th | $MgCl_2/TiCl_4$/Diester +TEA/Silane | 30-60** | 95-99 | Spherical catalyst, morphology control (spherical polymer particles with narrow PSD), high stereoselectivity |
| 5th | $MgCl_2/TiCl_4$/Diether+TEA | 70-120** | 95-99 | High activity, high stereoselectivity, spherical catalyst, morphology control (spherical polymer particles with narrow PSD) |

*Polymerization in hexane slurry at 70°C, $H_2$, 4h ** polymerization in bulk at 70°C, $H_2$, 2h

13

The majority of PP production processes today use heterogeneous ZN catalysts of $TiCl_4$ supported on $MgCl_2$ or $SiO_2$ with internal and external donors (4th and 5th generation ZN catalysts) because of their high activity, stereoselectivity and possibility to control the polymer morphology [5],[7].

## 2.1.2 Metallocene catalysts

Metallocene catalysts are generally transition metals which are coordinated with two organic ligands in a sandwich-like molecular structure. Most common for the olefin polymerization are transition metals of group IV of the periodic table of elements such as Ti, Zr and Hf. The ligands are cyclopentadienyl or cyclopentadienyl-derivative rings which can also be connected through bridges of different types (so called ansa-metallocene). Some examples of metallocene catalysts structures are shown in Figure 8. The structure of the ligands (bridged/ unbridged, bite angle, size, symmetry) influences the polymerization behavior enabling the production of polyolefins with different microstructures and characteristics [14],[38],[39],[40].

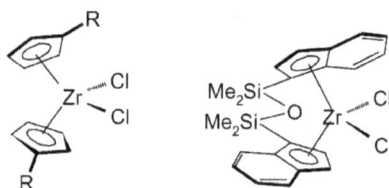

Figure 8: Examples for structures of metallocene catalysts for the olefin polymerization

Metallocene (MC) catalysts are very high active catalysts having per active center an activity 10 to 100 times higher compared to classical Ziegler-Natta catalysts [35]. In contrast to ZN catalysts, MC catalysts have only one sort of active site and are therefore also named as single site catalysts. They produce polymers with uniform properties with a narrow molecular weight distribution (PDI $\approx$ 2) and narrow chemical composition distribution [7],[14]. They are characterized by a high stereoselectivity producing polymer with high crystallinities but with a lower regioselectivity compared to ZN catalysts. Due to the regioerrors in the polymer chain, melting temperature of MC catalysts is lower. A comprehensive review on stereochemistry of MC catalysts is given by Brintzinger et al. [38] and Resconi et al. [39]. With metallocene catalyst produced polymers have an excellent transparency and good mechanical properties making them interesting for medical applications, food containers or bottles for personal care products [5]. In addition to its high activity and excellent microstructural control, metallocene catalysts could be adapted to existing olefin polymerization processes. They can be directly used in solution processes, whereas supported MC are used in slurry or gas-phase processes. A disadvantage is that MC catalysts are more expensive than Ziegler-Natta catalysts.

14

Furthermore, with one ZN catalyst type a wide product range can be produced, whereas with MC catalysts more specific resin types are produced [5],[7],[14].

First investigations of the polymerization with metallocene catalysts were carried out by Natta [41] and Breslow [42] already in the late 1950's. They show that a combination of titanocene dichloride ($Cp_2TiCl_2$) with aluminum alkyls (triethylaluminum or diethylaluminum chloride) can polymerize ethylene. The catalyst showed a poor activity and was not suitable to polymerize propylene. The early metallocene catalysts were mainly used as model catalysts in order to study the polymerization mechanism. The breakthrough of the MC catalysts, which made them interesting for the industrial usage, was derived by the development of Sinn and Kaminsky in the 1980's [43]. They synthesize methylaluminoxane (MAO) by the partial hydrolysis of trimethylaluminum (TMA) and used it as cocatalyst to activate and stabilize the metallocene catalyst. The activity of the developed catalyst system ($Cp_2ZrMe_2$/MAO) increased drastically, e.g. with one gram zirconium up to one ton of polyethylene could be produced. Brintzinger and co-workers [44],[45] synthesized ansa-metallocenes which allow a stereoselective polymerization. They were later used together with MAO by Kaminsky to polymerize highly isotactic PP [46].

In order to introduce the MC catalysts into existing polyolefin plants (slurry or gas-phase processes), they have to be supported on a solid, non-soluble material. Most common support for MC catalysts is silica ($SiO_2$) [14],[47],[48] but also other materials such as $MgCl_2$, alumina, zeolites or polystyrene are reported (see references 9 to 17 in [48]). There are different methods described in literature anchoring the MC catalyst and activator (MAO) on the support, see for example reviews from Hlatky [47],[49] and Fink et al. [48]. The main techniques can be classified into physical adsorption methods and covalent attachment methods [7]. In the first case, the order of immobilization of the MC and addition/activation with MAO can be varied. In the latter case, a functional group is added to the metallocene ligand which reacts with functional groups on the support surface. The different supporting methods (and also supporting conditions) result in different catalysts, thus, influencing catalyst activity as well as producing polymers with different properties. Also activity of supported MC catalysts is lower compared to the homogenous form and MWD of the polymer is broader [14],[48],[50].

## 2.2 (Kinetic) Modeling of olefin polymerization with coordination catalysts

### 2.2.1 Multi-scale modeling approach

For modeling of the polymerization processes different length scales have to be considered. In the leading work of Ray [51],[52], he proposed to classify the phenomena of the polymerization process into three length scale levels:

At the macro scale, phenomena on the reactor scale are considered such as overall heat and mass balances of the reactor, mixing, heat and mass transfer from and to the continuous phase, residence time distributions of the particles in the reactor and particle size distribution [53]. Macro- scale models are used for reactor design as well as for investigations of process control and stability.

At meso-scale, phenomena on the particle scale are considered. Catalyst fragmentation, particle morphology development, predictions of molecular weight distribution and chemical composition distribution, heat and mass transfer processes within the polymer particle (intraparticle) and between particle and continuous phase (interparticle) as well as phase equilibria are investigated [26],[54]. Different model approaches were developed in order to describe particle growth, formation of particle morphology as well as fragmentation behavior (see chapter 2.2.3).

At micro-scale, the molecular processes at the active sites of the catalyst are investigated. Models are used to describe the reaction kinetics (reaction rates of the polymerization process), nature of active sites, molecular property distributions, concentration of the reactants. Elementary reaction steps of the polymerization mechanism are considered. Depending on the modeling approach, kinetic models can be very complex or are reduced to simplified kinetic schemes describing phenomenological effects of the polymerization process (see chapter 2.2.2).

The three length scales of a polymerization process are connected to each other and models can become very complex. Often only a certain aspect of the polymerization is investigated in more detail. Therefore, models concentrate on describing the effects at a certain scale in detail while phenomena belonging to other scales are strongly simplified. A comprehensive summary of the multicomponent modeling of polymerization reactions in different reactor systems is given in the review of Dubé [54].

In this work, polymerization kinetics of different supported catalysts for the polymerization of propylene are investigated. The focus of the modeling is therefore set on the micro-scale level. As polymerizations were carried out under different reaction conditions (gas-phase and in bulk) as well as with and without prepolymerization also phenomena of the meso-scale need to be considered.

## 2.2.2  Kinetic schemes

Aim of the kinetic modeling is the mathematical description of the temporal course of the polymerization reaction in terms of rate constants and concentrations of the reactants in order to describe the catalyst behavior during the polymerization and to calculate molecular property distributions (e.g. molecular weight distributions).

Olefin polymerization with organometallic catalysts occurs via coordinative polymerization mechanism. Therein, the monomer is inserted between the polymerization active species (active site) of the catalyst and the growing polymer chain. The active species is formed by reaction between the transition metal species of the catalyst and the organometallic cocatalysts [55],[56],[57]. The most general accepted mechanism of chain growth is the monometallic mechanism according to Cossee and Arlman [58],[59]: In a first step, the C=C bond of the monomer is coordinated to the vacant site of the transition metal atom of the active species which is then inserted via a fourth center transition state into the metal-carbon bond (second step). The chain migrates to the position previously occupied by the coordinating monomer creating a new vacant site. Chain growth occurs until transfer reactions with another species (e.g. $H_2$, monomer, cocatalyst) or spontaneous (e.g. by ß-hydride elimination) lead to the formation of a dead polymer chain and a vacant polymerization active site which can initiate a further polymer chain [6],[7],[57].

The several elementary reaction steps involved in the polymerization mechanism with coordination catalysts are very complex (e.g. mechanisms of stereocontrol or copolymerization). For further reading it is referred to publications from Dusseult and Hsu [57], Kissin [56] and Albizetti et al. in [6] as well as Brinzinger [38] and Resconi et al. [39],[60] especially for polymerization mechanism with metallocene catalysts.

For the mathematical description of the polymerization reaction, phenomenological formal kinetic schemes are postulated based on experimental investigations. They consist of several reaction steps and can be more or less complex depending on the aim of the kinetic model, the process conditions (species involved) and the available experimental data used to estimate kinetic parameters. In literature, numerous publications can be found investigating the kinetics of propylene polymerizations with heterogeneous Ziegler-Natta catalysts [57],[61]-[67] as well as metallocene catalysts [68]-[73]. Reviews about kinetic modeling of olefin polymerizations are given by Xie et al. [74], Shaffer and Ray [75], Soares [76] and Touloupidis [77].

In general, a kinetic scheme for homopolymerization of olefins with coordination catalysts includes following main reaction steps:

1) Catalyst activation
2) Chain initiation
3) Chain propagation
4) Chain transfer
5) Site deactivation

Each reaction step can consist of a set of reactions depending on the involved reactants or detail of the kinetic mechanism. An example of a general kinetic scheme is given in Table 3.

Table 3: General simplified kinetic scheme for homopolymerization (based on [75])

| Description | Chemical equation | Rate constant |
|---|---|---|
| Catalyst activation | | |
| • by species A | $C_{pot} + A \rightarrow C^*$ | $k_{act,A}$ |
| • spontaneous | $C_{pot} \rightarrow C^*$ | $k_{act,sp}$ |
| | $A = CoCat, M, H_2$ | |
| Chain initiation | $C^* + M \rightarrow P_1^*$ | $k_i$ |
| Chain propagation | $P_n^* + M \rightarrow P_{n+1}^*$ | $k_p$ |
| Chain transfer | | |
| • to hydrogen | $P_n^* + H_2 \rightarrow C^* + D_n$ | $k_{tr,H2}$ |
| • to monomer | $P_n^* + M \rightarrow C^* + D_n$ | $k_{tr,M}$ |
| • to cocatalyst | $P_n^* + CoCat \rightarrow C^* + D_n$ | $k_{tr,CoCat}$ |
| • spontaneous | $P_n^* \rightarrow C^* + D_n$ | $k_{tr,\beta-H}$ |
| Site deactivation | | |
| • spontaneous | $P_n^* \rightarrow C_d + D_n$ | $k_{d,sp}$ |
| | $C^* \rightarrow C_d$ | $k_{d,sp}$ |
| • by species B | $P_n^* + B \rightarrow C_d + D_n$ | $k_{d,B}$ |
| | $C^* + B \rightarrow C_d$ | $k_{d,B}$ |
| | $B = Poison, CoCat, M, H_2, by-product$ | |

Catalyst activation:

In the activation step, the potential site of the catalyst ($C_{pot}$) is converted to a polymerization active vacant site ($C^*$). There are in principle four ways for the activation of the catalyst: by cocatalyst, by monomer, by hydrogen and spontaneously [55],[74],[75]. Usually, most approaches utilize only the activation by cocatalyst as this is the primary activation path.

Chain initiation:

In the chain initiation step, a monomer unit ($M$) reacts with the vacant active site producing the first living polymer chain with chain length one ($P_1^*$).

Chain propagation:

During chain propagation the polymer chain grows. A new monomer unit is inserted between the active site and the living polymer chain with chain length n ($P_n^*$) (see mechanism described above). The polymer chain increases by one monomer unit denoted as ($P_{n+1}^*$).

Chain transfer:

In the chain transfer reaction, the living polymer chain is terminated spontaneously (most often by β-hydride elimination) or by a further species acting as chain transfer agents forming a dead polymer with chain length n ($D_n$) and a vacant site [6],[10],[75]. The most common chain transfer agent is hydrogen which is used in industry to control the molecular weight [10]. Furthermore, chain transfers to monomer, to solvent and to cocatalyst are possible. The vacant site can further react starting with the chain initiation reaction.

Site deactivation

The activity loss during polymerization is generally described by the deactivation of the active sites (both vacant sites and active sites in the living polymer chain). They can undergo spontaneously deactivation or deactivate by reaction with another species, e.g. poison, by-product, monomer, cocatalyst or $H_2$ [57],[75].

In the kinetic scheme shown in Table 3, some simplifications were already made. Herein, only one sort of active site is assumed. In case of polymerization with single-site catalysts the given scheme can be used with a single set of the kinetic constants. In case of heterogeneous Ziegler-Natta catalysts it is generally assumed that two or more active site types are present, each one having its own set of kinetic constants creating polymer with different average properties (e.g. chain length, stereoregularity, comonomer compositions) [6],[26],[28],[54]. This multiple-site approach is for example used in order to model the broad MWD or CCD of polymer made with these catalysts [28],[54],[76],[78]. In the kinetic scheme each reaction step would be carried out for each assumed active site, e.g. see [75] and [76].

Another general kinetic scheme proposed by Shaffer and Ray [75] includes different chain end groups and also possible further reaction steps of site transformation as well as long chain branching (see also [70] and [76]).

In case of copolymerizations additional reactions with the comonomer have to be implemented where in the simple scheme only monomer reacts (during chain initiation, chain propagation, chain transfer) but also additional deactivation reactions of the chains with comonomer. Examples of copolymerization schemes are given by [75],[76],[78]-[80].

$H_2$ effect in kinetic models

From experimental studies it is known that $H_2$ influences the polymerization rate - in case of polymerization of propylene the polymerization rate increases [55],[81]. A most widely accepted mechanism is the dormant site theory, wherein dormant sites formed through the regioirregular insertion of the monomer can be reactivated by the reaction with hydrogen leading to further propagation (see chapter 2.3). For example in the studies of Pater et al. [62], Al-haj Ali et al. [82] and Samson et al. [83], the influence of $H_2$ on the kinetics of high active ZN catalysts in liquid propylene polymerization were investigated. Shaffer and Ray [75] investigated $H_2$ effect in olefin polymerization and implemented the $H_2$ effect in their general kinetic model. The proposed

19

kinetic schemes include additional reaction steps based on the dormant site theory. Following general steps are considered:

(1) *Formation of dormant sites:*

A dormant site ($Y_n$) is formed due to the regioirregular insertion of propylene into the growing chain.

$$(\text{I}) \qquad P_n^* + M \rightarrow Y_n$$

(2) *Reactivation of dormant sites by $H_2$:*

The dormant site is reactivated by chain transfer with $H_2$ forming an active site again and a dead polymer chain.

$$(\text{II}) \qquad Y_n + H_2 \rightarrow C^* + D_n$$

(3) *Reactivation of dormant sites by another species:*

A further reactivation of dormant sites by a species (S) is assumed which can be a monomer [62],[82] or a chain transfer agent other than $H_2$ [75],[83] or it can occur spontaneously [75]. Furthermore, Shaffer and Ray [75] assumed that after reactivation a vacant site and a living polymer are formed instead of dead polymer (no-death site transformation).

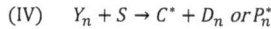

$$(\text{III}) \qquad Y_n + M \rightarrow P_{n+1}^*$$
$$(\text{IV}) \qquad Y_n + S \rightarrow C^* + D_n \; or \, P_n^*$$
$$(\text{V}) \qquad Y_n \rightarrow C^* + D_n \; or \, P_n^*$$

## 2.2.3 Meso scale particle modeling

When modeling polymerization reactions also phenomena at the polymer particle (meso scale), such as heat and mass transfer, catalyst fragmentation and particle growth, have to be considered as they can influence the reaction kinetics in terms of concentration of the reactants and temperature at the active sites.

Over the past decades, particle growth and morphology development have been studied intensively and different particle models were developed which differ in terms of catalyst/polymer particle morphology and in their description of inter- and intraparticle heat and mass transfer. In the following, only a very brief overview of the three main particle models will be given. Their basic concept in terms of particle morphology is schematically shown in Figure 9. Detailed descriptions of particle modeling can be found in the literature reviews of Dubé [54] and McKenna and Soares [26].

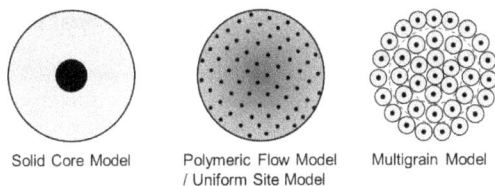

|  | Solid Core Model | Polymeric Flow Model / Uniform Site Model | Multigrain Model |

Figure 9: Schematic scheme of the main particle models for olefin polymerization with supported catalysts (black: catalyst/ catalyst fragments, grey: polymer) [26]

Solid-core model

The solid-core model (also core-shell model) is the first and simplest particle model and was proposed in early investigations by Begley [85] in 1966. Herein, it is assumed that the catalyst particle does not break up. The polymerization occurs at the catalyst surface, where all the active sites are located. During polymerization, the polymer grows in a shell-like layer of polymer around the solid catalyst core. Monomer is sorbed at the polymer surface and has to be transported through the polymer layer to the catalyst surface. This causes significant mass-transfer resistances, hence in industrial practice, solid core morphologies are avoided and porous catalysts fragmenting during polymerization are used. Hence, the solid core-model is only for theoretical considerations of interest.

Polymeric flow model

The polymeric flow model (PFM) was developed by Schmeal and Street [86], Singh and Merrill [87] and Galvan and Tirrell [88] in the 1970's and 1980's in order to describe the experimentally observed broad molecular weight distributions of polymer produced with ZN catalysts. The PFM assumes instantaneous catalyst fragmentation and treats the particle as a pseudo-homogeneous polymer matrix. Polymer formed by the reaction is transported outwards by convection. The resulting convective polymer flow also transports the catalyst fragments and leads to radial concentration gradients of the catalyst active components as well as the other species (monomer, chain transfer agent). The radial concentration profiles of catalysts/reactants lead to non-uniform polymer production within the particle and therefore it is able to describe broad molecular weight distributions. As the catalyst fragments and polymer phase are treated as one compact phase, transport processes of the reagents as well as heat are described by effective diffusion and heat transfer coefficients. Further examples using the PFM are given by Hoel et al. [89], Yiagopoulos [90] and Bartke [91].

A simplification of the PFM is the uniform site model. Herein, the catalysts particles are considered to be uniformly distributed in the polymer particle. Same concentration of active species over the whole polymer particle is assumed and can be therefore considered as locally independent. As the uniform site model is also considered as quasi-homogenous particle model, effective diffusion coefficients are used to describe transport processes. Examples are given by Patzlaff [92] and Sing and Merrill [87].

<u>Multigrain model</u>

The multigrain model (MGM) is one of the most widely used particle model in order to describe phenomena of growing particles in olefin polymerization. Early versions of the MGM were proposed by Yermacov et al. [93], Crabtree et al. [94] and Nagel et al. [95]. Further improvements of the MGM and detailed studies on the particle phenomena were carried out by Floyd, Hutchinson and Ray in the 1980's and 1990's [96] - [100]. As in the PFM, the MGM also considers instantaneous particle fragmentation at the beginning of the reaction but, in contrast, it takes the heterogeneous nature of the polymer particle into account. It is assumed that the polymer particle (macro particle or secondary particle) consists of an agglomerate of micro particles or primary particles. Each micro particle consists of a fragment of the original catalyst surrounded by the polymer layer according a core-shell morphology (described by the solid-core model). The reaction takes place at the active sites located on the surface of the catalyst fragment. Mass and heat transfer is therefore described on two different length scales: the mass-transfer through the pores of the macro particle (pore diffusion) and mass-transfer through the polymer layer to the active sites in the micro particles. For calculation of the monomer concentration at the active sites, two diffusion coefficients and diffusion lengths are necessary. An overview of the modeling equations for calculating monomer concentration and particle temperatures can be found in the reviews of Dubé et al. [54] and McKenna and Soares [26].

## 2.3 Role of hydrogen

Hydrogen is known as a very effective chain transfer agent in olefin polymerizations and is commonly used as molecular weight modifier in industrial practice [6],[7],[10]. A second effect of hydrogen is its influence on the rate of the polymerization reaction, which depends on the monomer and the nature of the catalyst [6],[81],[101]. Whereas in ethylene polymerization the presence of hydrogen generally leads to a decrease in the polymerization rate [55],[101]-[104], hydrogen leads to an increase of the polymerization rate in propylene polymerization [55],[61],[62],[82],[83],[101],[105]-[107].

The activating effect of hydrogen in propylene polymerization was observed by several researchers for (TiCl$_4$/MgCl$_2$ supported as well as non-supported) Ziegler-Natta catalysts, see further references 1-8 in [81] and [28]. Guastalla and Giannini [101] observed an increase of the polymerization rate of 150 % when adding hydrogen to the catalyst system MgCl$_2$/TiCl$_4$-AlEt$_3$. The effect of different external donors and hydrogen concentrations on the polymerization was investigated by Chadwick et al. [108]. The activating effect of hydrogen in propylene polymerizations was also observed with metallocene catalysts [73]. In contrast to the other catalysts (ZN and Phillips catalysts), much lower hydrogen concentrations are necessary for metallocene catalysts due to its high sensitivity toward hydrogen in terms of activity as well as molecular weight of the polymer [109].

The effect of hydrogen on the polymerization rate is reversible [61],[83]. The removal of hydrogen from the system leads to a decrease of the polymerization rate to its original level. When hydrogen is added again to the reactor, an immediate activation of the catalysts is observed.

Several hypotheses have been proposed explaining the rate enhancing effect of hydrogen in propylene polymerization. The most widely accepted explanation of the rate enhancing effect of hydrogen is its ability to reactivate dormant sites which are formed after a regioirregular 2,1 insertion of propylene into the growing chain [107],[108],[110]-[114]. Propylene can either be inserted into a growing polymer chain via a regioregular 1,2 insertion or a regioirregular 2,1 insertion, see Figure 10 (I). With the regioirregular 2,1 insertion or secondary insertion, respectively, the propylene molecule is inserted into the metal-carbon bond forming a dormant site which is relatively unreactive to further propagation (due to steric hindrance by the methyl group next to the Ti atom). This dormant site can be reactivated by chain transfer with hydrogen by ending the chain and releasing an active site for further propagation (Figure 10 (II)) [111].

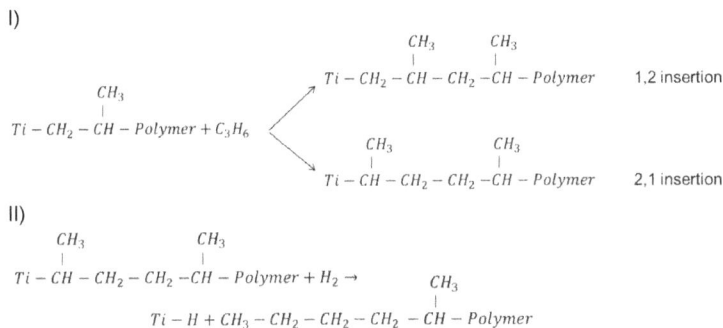

I)

$$Ti-CH_2-\overset{\underset{\displaystyle CH_3}{|}}{CH}-Polymer + C_3H_6$$

$$Ti-CH_2-\overset{\underset{\displaystyle CH_3}{|}}{CH}-CH_2-\overset{\underset{\displaystyle CH_3}{|}}{CH}-Polymer \qquad \text{1,2 insertion}$$

$$Ti-\overset{\underset{\displaystyle CH_3}{|}}{CH}-CH_2-CH_2-\overset{\underset{\displaystyle CH_3}{|}}{CH}-Polymer \qquad \text{2,1 insertion}$$

II)

$$Ti-\overset{\underset{\displaystyle CH_3}{|}}{CH}-CH_2-CH_2-\overset{\underset{\displaystyle CH_3}{|}}{CH}-Polymer + H_2 \rightarrow$$

$$Ti-H + CH_3-CH_2-CH_2-CH_2-\overset{\underset{\displaystyle CH_3}{|}}{CH}-Polymer$$

Figure 10: Formation and reactivation of dormant sites: I) Regioregular 1,2 and regioirregular 2,1 insertion of propylene, II) Chain transfer to hydrogen after 2,1 insertion [111]

A similar mechanism is proposed by Kissin et al. [81]. They considered the formation of an inactive center Ti-CH(CH_3)_2 which is formed by chain transfer with the monomer in a secondary 2,1 insertion or by a secondary insertion in the Ti-H bond formed by the chain transfer with hydrogen. The Ti-CH(CH_3)_2 species can be viewed as a dormant active center and is incapable to insert another propylene molecule. When $H_2$ is present, this species reacts with it and restores the Ti-H bond which can react further preferably with a primary propylene insertion.

23

## 2.4 Influence of prepolymerization step

Currently, coordination catalysts used for the production of polyolefins are highly active (supported) catalysts. Especially when the catalyst is injected in the process at high reaction temperatures and high monomer concentrations, a known problem is that catalyst particle overheating at the beginning of the polymerization can occur due to the high reaction rate conditions. Possibly consequences of particle overheating are an unwanted rapid thermal deactivation of the catalyst leading to a decrease of the overall polymerization rate but also the formation of agglomerates due to melting processes of the already formed polymer might occur [97],[98]. A second major effect is that the high reaction rate conditions can lead to an uncontrolled catalyst particle break up. Consequences are the loss of the (spherical) morphology of the resulting polymer particle and the formation of fines [115],[116]. Especially in continuous processes, the formation of fines can cause huge problems in the reactor or in the transition to the other stages, such as fouling, plugging or clogging.

In order to overcome these problems, a common solution in industry is the application of a prepolymerization step before the main polymerization, e.g. in the Borstar or Spheripol process (see chapter 1.3). In general, prepolymerization means that the polymerization starts at low rate conditions. Herein, the catalyst is injected at lower temperatures and lower monomer concentrations. The milder starting conditions have following main advantages: Firstly, particle overheating or thermal runaway of the catalyst particle in the initial stage of the reaction should be avoided due to the lower reaction rate conditions. Furthermore, the catalyst/polymer particle has time to grow providing a larger heat transfer area for the following main polymerization, thus, leading to less particle overheating. As a result, a higher overall polymerization rate is derived in the main polymerization and also the formation of agglomerates should be avoided. Secondly, fragmentation of the catalyst particles should occur in a more controlled way resulting in an improved polymer morphology and less fines. Thirdly, the low reaction rates should allow the particle to fully activate before it enters in the main polymerization process providing more active sites for the main polymerization [67],[116].

The effect of prepolymerization on activity and polymer morphology has been studied by several researches [62],[63],[67],[115]-[122]. In general, it was found that the application of a prepolymerization step leads to an increase of the polymerization rate during the main polymerization and thus to higher yields. Therein, it has been observed that prepolymerization temperature and time play a significant role. E.g. Samson et al. [67] showed that the yield could be increased by 20 to 30 % when applying a prepolymerization compared to polymerization without a prepolymerization step. Milder prepolymerization conditions and longer prepolymerization times were required when the main polymerization was carried out at higher temperatures.

The effect of prepolymerization on the resulting polymer morphology was investigated by Pater et al. [116]-[118]. They showed that the reaction rate conditions at the first stage of the polymerization determine the shape of the resulting polymer particle. At high reaction rate conditions, irregular shaped particles were formed and polymer powder of low bulk density was derived. Applying a prepolymerization led to very regular particles replicating the initial catalyst particle as well as higher bulk densities also at higher polymerization temperatures could be achieved.

Yiagopoulos et al. [90] investigated the single particle growth in the initial stage of polymerization and the effect of prepolymerization for a high active ZN catalyst in gas-phase olefin polymerization by using a modified polymeric flow model. They could show that the usage of right prepolymerization conditions reduces the internal and external mass and heat transfer resistances already in the initial stage of polymerization leading to reduction of particle overheating and enhancement of the polymerization rate. Zacca and Debling [123] modeled particle overheating phenomena and the effect of prepolymerization in different reactor configurations of olefin polymerization processes by using a population balance approach.

Nevertheless, despite the commercial importance of prepolymerization for production of polyolefins, only very limited quantitative data about the effect of prepolymerization on polymerization kinetics are available in open literature.

## 2.5 Kinetic measurements of (coordination) polymerization reactions under gas-phase and bulk conditions

Aim of kinetic measurements is to study the course of the polymerization reaction over the reaction time. Therein, the measuring principle depends on the polymerization process. In gas-phase or slurry polymerizations, kinetic information are accessible by measuring the feed of monomer to maintain a constant reactor pressure during polymerization (semi-batch operation of the reactor). In bulk polymerizations, where reactor pressure in principle corresponds to the vapor pressure of the liquid monomer, the semi-batch approach is not applicable. In that case, calorimetry or dilatometry can be used. In the following, the main principles for the online measurement of the polymerization rate will be presented.

### 2.5.1 Kinetic measurements of gas-phase and slurry polymerizations

For the kinetic investigation of the polymerization in laboratory scale, stirred tank reactors or autoclaves with typical reactor volumes between 0.1 to 5 l are used. They are equipped with a cooling unit (cooling jacket and/or thermostat bath) in order to remove the polymerization heat as well as with pressure sensors in order to control the reactor pressure. In general, during polymerization, gaseous monomer is converted by means of the coordination catalyst into solid polymer particles. Under batch conditions, where no educts are fed and no products are removed, the monomer conversion would lead to a pressure drop in the reactor as the density of the polymer is higher than the density of the gaseous monomer. In semi-batch operation, monomer is fed into the reactor in order to compensate this pressure drop. The feed of monomer is controlled by a pressure control loop keeping the pressure constant. At isothermal conditions, the fed amount of monomer, which is measured by a flow meter, corresponds to the monomer consumption and is proportional to the gross reaction rate of the polymerization. By relating the consumed monomer to the amount of catalyst, the current activity of the catalyst can be determined at any reaction time leading to the catalyst specific activity-time profiles (see chapter 4.6.1).

For kinetic measurements it is important to maintain the reaction at isothermal conditions as changes of the pressure with changing temperatures would influence the pressure control. In order to obtain reliable kinetic data from the monomer consumption, mass transport limitations of the monomer (from gas-phase to polymer phase) need to be avoided. When monomer consumption does not increase with increasing stirring speed, external mass transport limitations can be excluded [124].

This method is well established in literature and examples for kinetic measurements of coordinative catalysts for the gas-phase polymerization of propylene can be found in the investigations of Choi and Ray [61], Soares and Hamielec [28], Samson et al. [63], Meier, Weickert and van Swaaij [64, 73] and Piddhun [125]. Kröner, T. [80] describes the kinetic investigation of heterophasic copolymerization of propylene with ethylene using the same measuring approach. Herein, the comonomer was fed in a constant ratio to the monomer. By measuring the gas compositions, the individual consumption rates of the monomers could be monitored. Furthermore, kinetics of coordinative catalysts for olefin polymerization in slurry can be investigated based on the same measuring principle as for gas-phase polymerizations. Examples can be found in the investigations of Keii [65], Kahrmann [66] or Kröner, S. [124].

### 2.5.2 Kinetic measurements of bulk polymerizations

In bulk polymerizations, the reaction is carried out in liquid monomer. When only minor partial pressures of other components (e.g. $H_2$ or $N_2$) are present, the reactor pressure nearly correspondents to the vapor pressure above the liquid monomer which remains constant with conversion. The only kinetic information per experiment is the yield derived at the end of the reaction. In order to determine a kinetic profile, a set of experiments at different reaction times with a good reproducibility would be necessary. In literature, only few references about the online measurement of the kinetics of the bulk polymerizations of propylene can be found using either reaction calorimetry or dilatometry.

The dilatometric method uses the shrinkage of the reactors content due to differences in density of monomer and polymer. In a fully filled reactor, the volume shrinkage will lead to a sensitive decrease in reactor pressure which is used as kinetic signal. Al-haj Ali et al. [126] investigated the bulk polymerization with a ZN catalyst in fully filled reactor. For the estimation of the reaction rate they implemented a functional equation which uses the compressibility and volume expansivity of the reaction mixture as a function of pressure and temperature. The method gave comparable results with calorimetric measurements. However, not all parameters for the dilatometric method were known accurately and e.g. the effect of $H_2$ was not considered in this study. Another study using the dilatometric approach was carried out by Patzlaff [92]. He investigated the kinetics of different ZN catalysts in bulk phase polymerization in a partially filled reactor. Therein, the catalyst was injected with pentane. With consumption of monomer, the composition of the liquid mixture (monomer and pentane) changed resulting in a significant effect on the vapor pressure. The polymerization rate was determined from the measured pressure drop taking into account the vapor-liquid equilibrium of the reaction mixture. Hence, accuracy of the method depends largely on the quality of the used vapor-liquid equilibrium data.

A further method suitable to investigate kinetics of bulk polymerizations is reaction calorimetry. Herein, the reaction kinetic is determined by measuring the heat generated by the exothermal polymerization reaction. The released reaction heat is equal to the rate of polymerization times the reaction enthalpy according:

$$\dot{Q}_{Chem} = R_P \cdot V_R \cdot (-\Delta H_R) \qquad [W] \tag{2.1}$$

wherein $\dot{Q}_{Chem}$ (W) is the released heat by the chemical reaction and $\Delta H_R$ (J/mol) is the reaction enthalpy, which is negative for exothermal reactions.

In general, a reaction calorimeter consists of a stirred reaction vessel surrounded by a cooling jacked with a circulating thermostating liquid, which removes the generated heat to a cooling system. Calorimetric methods can be classified by their principle of heat measurement (e.g. heat flow, heat balance or power compensation calorimetry) and their mode of temperature control (isothermal, adiabatic, isoperibole). Zogg et al. [127] reviews the most important calorimetric principles for kinetic analysis of chemical reactions in detail.

A most often used calorimetric method for polymerization reactions is the isothermal heat flow reaction calorimetry. Herein, the heat flow from the reactor content through the reactor wall into the thermostating liquid is determined by measuring the temperature difference between reactor content and thermostating liquid. The heat flow is calculated according the heat transfer equation:

$$\dot{Q}_{HF} = k \cdot A \cdot (T_R - T_J) \qquad [\text{W}] \qquad (2.2)$$

wherein $\dot{Q}_{HF}$ is the removed heat flow (W), $k$ is the overall heat transfer coefficient (W/(m²·K)), $A$ is the overall heat transfer area (m²), $T_R$ is the reactor temperature and $T_J$ is the mean jacket temperature (K).

Main drawback of the method is that especially for polymerization reactions $k$ and often also $A$ are not constant. Typical phenomena during polymerization like increasing viscosity of the reaction mixture, changing composition of the reaction mixture or fouling at the reactor walls can influence the heat transfer coefficient. Also the filling level and therefore the active heat transfer area can change during polymerization due to different densities of monomer and polymer.

Samson et al. [67] investigated the kinetics of a highly active ZN catalyst under bulk conditions in liquid propylene using heat flow calorimetry.

Korber et al. [128],[129] studied the kinetics of a silica supported metallocene/MAO catalyst in propylene bulk phase as well as in propylene slurry polymerization using a RC1 reaction calorimeter. The calorimeter is based on the heat flow technology measuring the temperature difference between reactor and jacked, the term of heat transfer coefficient and heat transfer area is determined by a calibration heater.

Further studies of the bulk polymerization of propylene using the (isothermal) heat flow calorimetry are given by Meier et al. [73] and Pater et al. [62],[117],[118].

An example of isoperibolic heat flow calorimetry is given by Al-haj Ali et al. [126]. Herein, jacket temperature is kept constant during the reaction and the temperature difference between the cooling jacked and the reaction mass is related to the polymerization rate. A drawback of this method is that for accurate measurements the temperature increases should be low. Therefore, only small amounts of catalysts can be used which would increase the probability of catalyst poisoning [124],[126].

In this work, a special reaction calorimeter by ChemiSens® (CPA202) was used working based on the principle of heat flow calorimetry [130]. Herein, the heat flow can only pass through the reactor base which is equipped with temperature sensors placed in a defined distance in the base and a Peltier element acting as reversible heat pump. The reactor walls are active insulated. The whole reactor is placed in a thermostating bath and is operated at isothermal conditions. The main advantage of this calorimeter is that the heat flow can be measured directly via the heat conductivity in the reactor base:

$$\dot{Q}_{HF} = \lambda \cdot \frac{A}{d}(T_1 - T_2) \qquad [\text{W}] \qquad (2.3)$$

where $\lambda$ is the specific heat conductivity coefficient of steel (W/(m·K)), $A$ is the defined area of the reactor base (m²), $d$ is the distance between the temperature sensors (m) and $T_1$- $T_2$ is the measured temperature difference in the reactor base (K).

The measurement is independent of the heat transfer conditions and filling level in the reactor and therefore no calibration of heat transfer coefficients like for conventional heat flow method is needed.

# 3 Aim of the investigations and thesis outline

Due to the economic importance, the development of improved coordination catalysts for the industrial production of polypropylene is continuously ongoing. Since in coordinative polymerization all kinetic constants are catalyst dependent, the continuous development of new catalysts leads to a continuous need for kinetic measurements under industrially relevant conditions. Due to continuous improvement in catalyst activities, the role of prepolymerization in order to avoid catalyst overheating is increasing.

Aim of this work is to investigate the kinetics of different supported coordination catalysts for the polymerization of propylene under industrially relevant polymerization conditions. Therein, two $4^{th}$ generation Ziegler-Natta catalysts of different activity are investigated under gas-phase conditions and a supported metallocene catalyst is studied under bulk conditions.

Catalyst kinetics are studied as function of operation conditions such as temperature, pressure and hydrogen concentration. A particular focus is set on the investigation of the effect of catalyst injection conditions (respectively prepolymerization) on catalyst kinetics.

Two different experimental setups were used in order to measure the kinetics of the polymerization at the different reaction conditions: For the gas-phase polymerization, a 5 l horizontal stirred tank reactor working at semi-batch conditions was used. The bulk polymerization was carried out in a 250 ml reaction calorimeter which works based on the heat flow calorimetry. The produced polymer powders were further analyzed using different analytical methods in order to characterize polymer properties as well as polymer morphology.

Furthermore, a simplified phenomenological kinetic model approach is developed for each catalyst type which shall be able to describe the kinetic behavior at the different reaction conditions and to determine the kinetic constants.

The work is divided into two parts according to the different catalyst types and reaction conditions: In the first part, the gas-phase polymerizations with the two Ziegler-Natta catalysts as well as the model development are described. In the second part, the experimental investigations and the modeling of the propylene bulk polymerization with a supported metallocene catalyst are presented.

# 4 Experimental investigation of the gas-phase polymerization of propylene with two different Ziegler-Natta catalysts

## 4.1 Experimental setup for the gas-phase polymerizations

The experiments were carried out in an existing laboratory setup which enables the kinetic investigation of gas-phase polymerizations of olefins [80],[125],[131]. A scheme of the adopted experimental setup is shown in Figure 11. In general, it consists of four main parts: The raw material supply with monomer purification system, the polymerization reactor, the controlling unit and the recording unit.

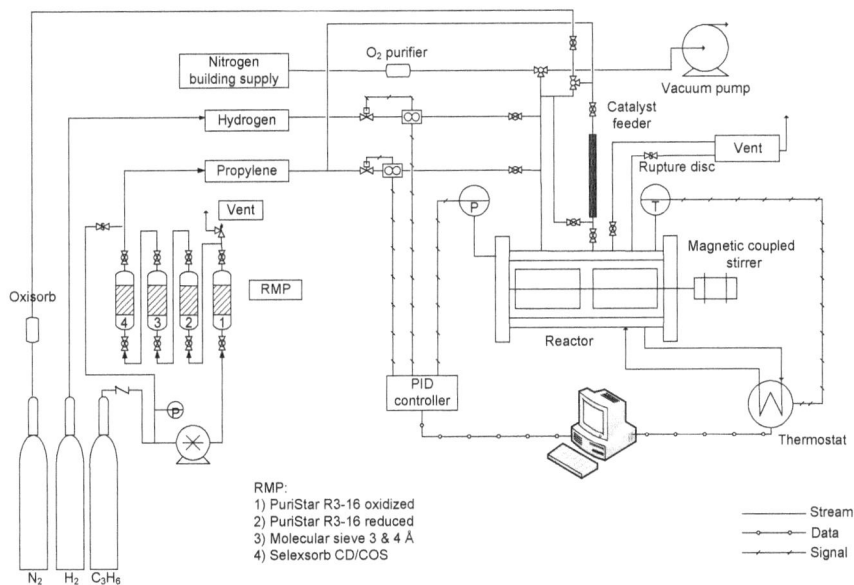

Figure 11: Scheme experimental setup for the gas-phase polymerization of propylene

## 4.1.1 Raw material supply and monomer purification system

For the homopolymerization reaction, liquid propylene as monomer, hydrogen as chain transfer agent and high pressure nitrogen (e.g. for catalyst injection) were used. All gases were supplied in gas bottles. The grades of the gases are summarized in Table 4. Furthermore, low pressure nitrogen of high purity was used for inertization of the reaction system from the building supply (liquid nitrogen storage).

Table 4: Raw materials and purity

| Gas | Supplier | Purity |
|---|---|---|
| Propylene | Linde/ Air Liquide | 2.5 |
| Hydrogen | Linde | 5.0 |
| Nitrogen (high pressure) | Linde | 6.0 |
| Nitrogen (low pressure) | Air Liquide | high grade |

The used Ziegler-Natta catalysts are very sensitive to water and oxygen and other polar compounds. Despite the high purities of the educts, an additional four-stage purification system for the monomer was installed in order to ensure high catalyst activities and good reproducibility [80]. In the first column, an oxidized copper oxide catalyst (PuriStar® R3-16, BASF) was used to oxidize CO to $CO_2$. In the second column, the same copper oxide catalyst but in reduced form was used in order to remove oxygen. In the third stage, the column was filled with molecular sieve of pore size 3 to 4 Å removing traces of water. In the last column, a selective adsorbent (Selexsorb COS, BASF) was used to remove $CO_2$ and sulfur compounds and last traces of water. For the nitrogen, an oxisorb-cartridge (Linde) was used to remove possible solved oxygen. Hydrogen was used without any further purification.

Before entering the purification system, the liquid propylene was pressurized up to 40 bars in order to enable liquid feeding to the reactor. In order to ensure a good purification, the liquid propylene was cycled several times through the purification system ensuring high contact between liquid monomer and purification catalysts.

## 4.1.2 Lab-scale polymerization reactor

The polymerizations were carried out in a 5 l horizontal stirred tank reactor which is specially constructed in order to carry out polymerizations under industrial relevant conditions. Pressures up to 40 bar and temperatures up to 120 °C can be realized. A picture of the reactor is shown in Figure 12. The stainless steel reactor is heated or cooled by a double jacket cooling around the cylindrical shell of the reactor. In order to maintain the temperature, a thermostat (Proline RP 855, Lauda) was used having a maximum heating power of 3.5 kW and maximum cooling power of 1.6 kW. The temperature inside the reactor is measured with a residence thermometer (Pt 100) placed through the cooling jacket in the middle of the reactor. The reactor pressure is measured by a pressure sensor from WIKA (IUT-10, WIKA) within a range from 0 to 40 bar with an error of 0.15 %. The reactor is equipped with a magnetic coupled horizontal stirrer driven by a Büchi three phase variable gear motor. Stirring speeds in the range from 0 to 800 rpm are possible to maintain. The horizontal position of the reactor has two main advantages:

Firstly, it improves the heat removal and, secondly, a better mixing comportment compared to vertically stirred reactors can be realized [125].

The liquid propylene, hydrogen and catalyst are inserted from the top of the reactor. The liquid propylene is fed via a liquid phase mass flow controller (Flomega 5882, Brooks), whereas hydrogen is fed via a gas-phase mass flow controller (MF50S, Brooks). The mass flow controllers are connected to a control box (brooks instruments), see next chapter 4.1.3.

In order to inertize the reactor and the piping system, nitrogen and a two-stage rotary vane vacuum pump (P 6 Z, Ilmvac) is connected. A detailed description of the cleaning procedure will be given in chapter 4.3.1.

Figure 12: Picture reactor setup for the gas-phase polymerization

## 4.1.3 Measuring principle and control units

The reactor works in semi-batch mode in order to measure the reaction kinetics online (see chapter 2.5.1). Therein, monomer is fed into the reactor in a closed pressure control loop maintaining a constant reactor pressure. Under isobaric and isothermal conditions, the fed amount of monomer corresponds to the monomer consumption during reaction which is proportional to the gross reaction rate of the polymerization. Figure 13 shows a scheme of the pressure control loop. As previously described, the reactor pressure is measured by a pressure sensor (IUT-10, WIKA). The pressure value is transmitted to a PID controller (i16, Omega i-Series, Newport) where it is compared with a given pressure set point. Depending on the pressure difference between given and set value, the PID controller sends a signal to the control box (Brooks instruments) which regulates the mass flow rate of propylene until constant reactor pressure is reached.

In order to determine the kinetics, isothermal conditions need to be maintained throughout the whole reaction. For that purpose, the thermostat (Proline RP 855, Lauda, see section 4.1.2) was used which regulates the reactor temperature by a cascade closed loop control. The reaction temperature is measured by the resistance thermometer (Pt 100) and compared with the set point temperature at the thermostat. In the "external" mode, the bath temperature is adjusted in that way that the reactor temperature remains constant. As thermostating liquid, a thermal oil (Fragol Therminol® ADX 10) was used circulating through the double jacket of the reactor. The temperature control loop is shown schematically in Figure 13.

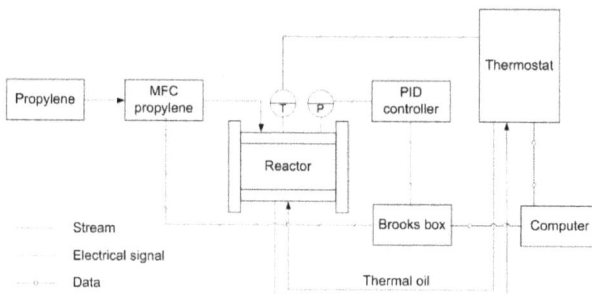

Figure 13: Scheme control loop of the pressure and temperature regulation

### 4.1.4 Data acquisition system

For the data acquisition the software DASYLab9 (National instruments) was used. A snapshot of the user interface is shown in Figure 14. With the software the measured current mass flows, pressures and temperatures are collected, processed and recorded. In the flow sheet, the current process parameters are visualized by digital displays and saved. For safety reasons, process inputs such as reaction temperature and reaction pressure can be also set manually directly at their control units.

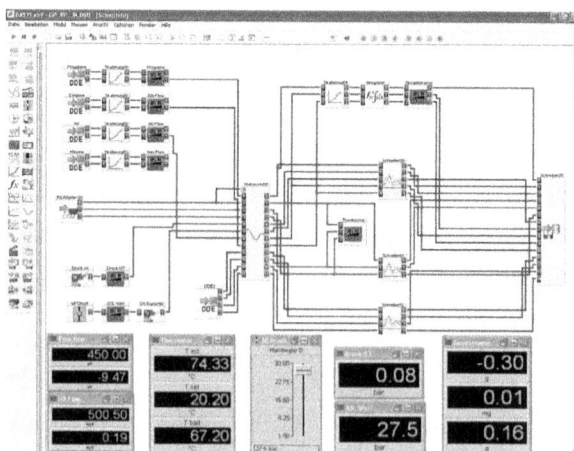

Figure 14: Data acquisition user interface DASYLab9

34

## 4.2 Catalysts and chemicals – handling and preparation

For the investigation of the gas-phase polymerization two different supported Ziegler-Natta catalysts of the 4th generation were provided by the industrial cooperation partner. Selected characteristics are summarized in Table 5. The active component of both catalysts is titanium. The catalyst particles are spherical having a narrow particle size distribution. In general, catalyst B has a higher activity compared to catalyst A.

Table 5: Selected characteristics of the investigated Ziegler-Natta catalysts

| Property | Unit | Cat A | Cat B |
|---|---|---|---|
| Active component | | Ti | Ti |
| D50 value | µm | 56.4 | 55.0 |
| Density | g/l | 2650 | 2320 |

As cocatalyst a 93 % solution of triethylaluminum (TEA, Sigma Aldrich) was used. Beside the function as cocatalyst, TEA also acts as scavenger by reacting with impurities inside the reactor. Therefore, a fix amount of 0.45 ml TEA was used in all polymerization reactions. As external donor a 0.1 molar solution of cyclohexyl-dimethoxy-methylsilane in heptane (both from Sigma Aldrich) was used. Donor was used in a fixed ratio to the weighted catalyst amount (fixed Do/Ti ratio). Liquid propylene was used as monomer and hydrogen as chain transfer agent (see Table 4, chapter 4.1.1 ).

TEA is highly reactive in contact with air or water. Also the catalyst is deactivated by the contact to air or other impurities. Therefore, the preparation of the catalyst, cocatalyst and donor was carried out in a glove-box (Jacomex) under inert conditions. A special double-feeder construction was used in order to avoid any pre-contact between catalyst and cocatalyst before the reaction, see Figure 15. In one chamber the pure catalyst was filled together with 1 ml heptane. In the second chamber the cocatalyst and the external donor were inserted. The catalyst-heptane slurry was chosen for two main reasons: Firstly, the catalyst particles are very small so that due to static loading some particles can be lost or are not inserted into the reactor exactly. In addition, evaporation of heptane during injection will cool the catalyst and reduce overheating.

Figure 15: Scheme catalyst - cocatalyst injection feeder

## 4.3 Polymerization procedures

### 4.3.1 General reactor preparation

Prior to every experiment, the reactor as well as the piping system was inertized. For that purpose, the reactor was heated up to 100°C to evaporate any moisture on reactor walls. Then vacuum was applied for approx. 30 min followed by flushing the reactor with $N_2$. This alternating vacuum - $N_2$ flush cycle was repeated minimum for six times. At the end, the reactor was left under $N_2$ overpressure. When an experiment was started, $N_2$ was released and the reactor was prepared as described in the specific procedures in the following sections.

Catalyst, cocatalyst and donor were prepared in the glove-box directly before the reaction start as described in section 4.2. The feeder was connected to the reactor system under $N_2$ flow and inertized with alternating vacuum and $N_2$ flush. The catalyst and cocatalyst/donor were injected by means of liquid propylene. The reaction started directly with the catalyst injection.

### 4.3.2 Polymerization without prepolymerization step

In order to generate kinetic data close to reaction conditions in commercial plants, a first polymerization procedure without a prepolymerization step was developed. Herein, the catalyst was injected directly under reaction conditions. Different catalyst/cocatalyst injection procedures were developed in order to find a reliable and reproducible polymerization procedure. The following conclusions could be made from preliminary performed studies:

- the catalyst should be added before cocatalyst for better distribution in the reactor
- the time lag between injection of catalyst and cocatalyst should be short
- the addition of any scavenger solution prior to catalyst injection is not necessary
- the catalyst injection should be carried out with liquid propylene instead of high pressure nitrogen (promotes agglomeration).

The final polymerization procedure is shown schematically in Figure 16. The inertized and evacuated reactor was filled with propylene and hydrogen and heated up to the desired reaction temperature. Catalyst and cocatalyst together with donor were injected directly at reaction conditions and the polymerization was carried out for one hour with a stirring speed of 350 rpm. The reaction was stopped by pressure release and cooling down the reactor. The reactor was further flushed with compressed air in order to fully deactivate the catalyst and cocatalyst before opening the reactor and removing the polymer.

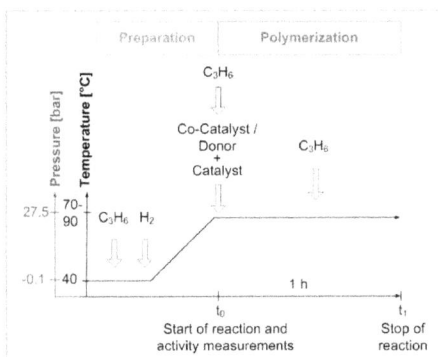

Figure 16: Gas-phase polymerization procedure 1: Polymerization without prepolymerization step

## 4.3.3 Polymerization with prepolymerization step

In the second procedure, polymerizations were carried out with a prepolymerization step. Herein, the catalyst was injected under mild reaction conditions (low reaction temperature, low pressure and therefore low monomer concentration). A scheme of the polymerization procedure is shown in Figure 17. After inertization, propylene and hydrogen were fed into the evacuated reactor. The reactor was heated up to the initial temperature of 40 °C (partial pressure of propylene 13 bar). The catalyst, cocatalyst and donor were injected with liquid propylene and the prepolymerization started. During the prepolymerization, which was carried out for approx. 15 min, the reactor was heated up and filled with propylene until the desired reaction conditions were reached. It has to be noticed that reactor filling was carried out slowly in order to maintain gas-phase conditions. When the main polymerization conditions were reached, the reaction was carried out for one hour. The stirring speed during the reaction was set equal to 350 rpm. The polymerization was stopped as described before.

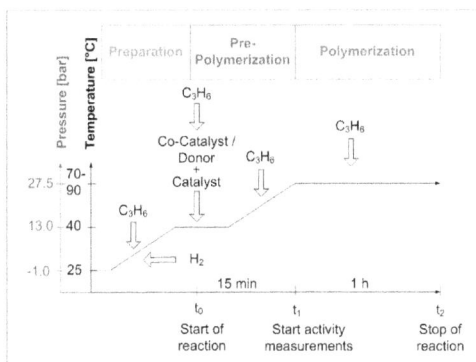

Figure 17: Gas-phase polymerization procedure 2: Polymerization with prepolymerization step

## 4.4 Reaction conditions and experimental plan

In the experimental study, the influence of reaction temperature, pressure, hydrogen concentration as well as different catalyst injection conditions (with and without prepolymerization step) on the catalyst activity were investigated for the two ZN catalysts A and B. The polymerizations were carried out within the following industrial relevant operating window:

- Reaction temperature: 50 to 90°C
- Pressure: 25 to 30 bar
- $H_2$: 0 to 500 mg (corresponding to 0 - 0.05 mol/l)
- Prepolymerization: $T_{cat-injection} = 40°C$

As described in chapter 4.2, a fixed amount of TEA (0.45 ml) was used as cocatalyst for both catalysts. The amount of the donor was related to the catalyst amount and is catalyst specific. The detailed experimental plans for catalyst A and catalyst B are summarized in Table 6 and Table 7, respectively. At the lower reaction temperatures of 50°C and 60°C, the corresponding vapor pressure of propylene is 20.26 bar and 24.73 bar, respectively (NIST Standard Reference database [132]). Hence, the propylene reaction pressure had to be reduced in order to avoid condensation and ensure gas-phase conditions. For a better comparability, the propylene partial pressure for the low temperature experiments was adjusted according to the same equilibrium monomer concentration in the polymer ($c_M*$) as for the experiment at 80°C and 27.5 bar propylene. $c_M*$ was calculated using Henry's law and Stern equation (see chapter 5.1.8). At 80°C and 27.5 bar $c_M$ is 1.617 mol/l. The corresponding partial pressure at 50°C is 16.4 bar and at 60°C 19.8 bar.

Table 6: Experimental plan gas-phase polymerization with ZN catalyst A

| Influence | Temperature | Pressure | Hydrogen | Si/Ti |
|---|---|---|---|---|
| Unit | °C | bar | mol/l | mol/mol |
| Pressure | 80 | 25.0; 27.5; 30.0 | 0.025 | 3.1 |
| Hydrogen | 70; 80; 90 | 27.5 | 0; 0.0025; 0.01; 0.025; 0.05 | 4.0 |
| Temperature | 50*; 60** | 16.4*/19.8** | 0.025 | 4.0 |
| | 70; 80; 90 | 27.5 | 0.025 | 4.0 |
| Prepolymerization | 70; 80; 90 | 27.5 | 0.025 | 4.0 |

| Influence | Temperature | Pressure | Hydrogen | Si/Ti |
|---|---|---|---|---|
| Unit | °C | bar | mol/l | mol/mol |
| Pressure | 70 | 25.0; 27.5; 30.0 | 0.025 | 23 |
| Hydrogen | 70; 80; 90 | 27.5 | 0; 0.0025; 0.01; 0.025; 0.05 | 23 |
| Temperature | 50*; 60** | 16.4*/19.8** | 0.025 | 23 |
| | 70; 80; 90 | 27.5 | 0.025 | 23 |
| Prepolymerization | 60; 70; 80; 90 | 27.5 | 0.025 | 23 |

*/** Reduced pressure corresponding to an equilibrium monomer concentration within the polymer particle of $c_M$* = 1.617 mol/l (reference point: T = 80 °C, p = 27.5 bar)

## 4.5 Analytical methods

### 4.5.1 Melt mass flow rate

The melt mass flow rate (MFR) was measured in order to determine the weight average molecular weight ($M_w$) of the produced polymer samples. The MFR is defined as mass of molten polymer in grams flowing through a capillary of a specific diameter and length at a polymer specific pressure (applied by gravimetric weights) and temperature within ten minutes. The measurements were carried out by the industrial cooperation partner with the Melt FloW@on (Emmeram Karg Industrietechnik) according the MFR test standards ASTM D 1238 [133] and ISO 1133 [134], respectively. The parameters for polypropylene are summarized in Table 8. In addition, the MFR of selected polymer samples was measured in our own laboratory with a micro flow melt indexer (CSI-127MF, Custom Scientific Instruments, Inc.). From the derived MFR values the weight average molecular weights were estimated using a MFR - $M_w$ correlation provided by the industrial partner (equation 4.1), wherein the $M_w$ of polymer samples of known MFR value were measured via GPC.

$$M_w = 538445 \cdot MFR^{-0.249} \qquad \left[\frac{g}{mol}\right] \qquad (4.1)$$

Table 8: Parameters for melt flow rate measurement of polypropylene samples according standards ASTM D 1238 [133] and ISO 1133 [134]

| Parameter | Value | Unit |
|---|---|---|
| Test temperature | 230 | °C |
| Test weight | 2.16 | kg |
| Pre-heating time | 300 | s |

### 4.5.2 Crystallinity

The crystallinity of the semi-crystalline polymer samples was determined by using differential scanning calorimetry (DSC) carried out by the industrial cooperation partner. The measurements were carried out with the polymer powder in order to determine the properties close to the state during polymerization. The measurements were performed under nitrogen flow with a standardized temperature program. A typical curve of a DSC measurement is shown in Figure 96 in appendix 9.1.2.1. The crystallinity of the polymer sample was determined from the specific melt enthalpy of the first melting peak (mean value of two DSC measurements) and the specific melt enthalpy of a 100 % crystalline polypropylene. The first melting peak was chosen in order to determine the crystallinity of the polymer product close to the state directly after the polymerization. The crystallinity K was calculated according following equation:

$$K = \frac{\Delta H_m}{\Delta H_{m,100\%}} \cdot 100\% \qquad [\%] \qquad (4.2)$$

with $\Delta H_m$ as measured specific melt enthalpy from the polymer sample and $\Delta H_{m,100\%}$ as specific melt enthalpy of an ideal crystalline polypropylene. For polypropylene $\Delta H_{m,100\%}$ is 207 J/g [135].

### 4.5.3 Porosity

Porosity measurements of the polymer samples were carried out at the Department of Industrial Chemistry at the Martin-Luther-University Halle-Wittenberg. Two measurements were performed successively: firstly at low pressures (up to 400 kPa) with the porosimeter Pascal 140 (Thermo Finnigan) and afterwards at high pressures (up to 400 MPa) with the porosimeter Pascal 440 (Thermo Finnigan). Mercury was used as intrusion medium. As result, the total porosity was derived by combining the results of both measurements with the provided software Pascal.

### 4.5.4 Bulk density

The bulk density of the polymer is defined as mass of a packed bed of polymer particles within a certain volume. The total volume includes the pore volume inside the polymer particle and the volume between the particles. The bulk density is influenced by particle morphology, porosity and particle size distribution. The measurement of the bulk density was carried out gravimetrically by weighting a loose packed bad of the polymer sample. A measuring cylinder of a defined volume was totally filled with the polymer sample and the polymer mass was weighted by means of a balance. The bulk density was than calculated according following equation:

$$\rho_{Bulk} = \frac{m_{PP}}{V_{cylinder}} \qquad \left[\frac{g}{l}\right] \qquad (4.3)$$

## 4.5.5 Particle size distribution

The particle size distribution (PSD) of the polymer powder was determined by sieving analysis carried out externally by the cooperation partner as well as at the Center of Engineering, Professorship of Thermal Process Technology at the Martin-Luther-University Halle-Wittenberg. Therein, the weight fraction of a particle size fraction was derived from the weighted mass of the sieved polymer fraction divided by the total mass of all polymer particles, see equation 4.4. The $D_{50}$ value or median of the distribution was derived from the cumulative frequency distribution. The $D_{50}$ value is defined as the particle size where 50% of the population have particle sizes above that value and 50% have particle sizes below that value, respectively.

$$w_{PP,i} = \frac{m_{PP,i}}{\sum_i m_{PP,i}} \cdot 100 \qquad [wt\%] \qquad (4.4)$$

Furthermore, PSD was measured by the external cooperation partner using Malvern analysis (Mastersizer, Malvern Instruments Ltd.). It has to be noticed that only polymer sample with a particle size below 2000 μm could be analyzed. Therefore, the particle fraction above 2000 μm was sieved out before the measurement.

## 4.5.6 Scanning electron microscopy

The morphology of the polymer particles was analyzed via scanning electron microscopy (SEM) at the research group General Material Science at the Martin-Luther-University Halle-Wittenberg. The polymer samples were fixed on a SEM sample holder and sputtered with a conductive gold layer under vacuum with the sputter coater S150 B from Edwards (90 s, 40 mA). The prepared samples were analyzed with the scanning electron microscope JSM 6300 (JOEL) under high vacuum ($5 \cdot 10^{-5}$ mbar) with an acceleration voltage of 10 kV in the secondary electron mode. The images were edited and recorded with the software AnalySIS.

## 4.5.7 Sorption measurements

In order to study the solubility of the monomer into the polymer particles of different morphology, gravimetric adsorption measurements were performed with a magnetic suspension balance (Rubotherm GmbH) at our laboratory. The sorption measurements were carried out with untreated polymer particles at 70°C and a monomer (propylene) pressure of 7 to 8 bar. A comprehensive description of the setup and the performance of sorption measurements are given in the work of Kröner [80].

## 4.6 Results kinetic measurements of the gas-phase polymerization with two different Ziegler-Natta catalysts

In the following chapter, the experimental results of the gas-phase polymerization are presented. The influence of the different reaction conditions on catalyst activity, the kinetic profiles and the polymer properties are discussed for both ZN catalysts. The conclusions from experimental results are the basis for the derivation of the kinetic model and are presented in the next chapter 5.

### 4.6.1 Gained experimental data and calculation of activities

As described in section 2.5.1, kinetic information in gas-phase polymerizations are accessible via measuring the mass flow of propylene. At constant reaction temperature and pressure, the measured mass flow of propylene corresponds to the current consumption of the monomer by the polymerization reaction. The monomer consumption is proportional to the gross reaction rate of polymerization or the current activity of the catalyst, respectively. The current activity is defined as the amount of produced polymer per amount catalyst and time and is determined from the measured mass flow of propylene according following equation:

$$A = \frac{m_{PP}}{m_{Cat} \cdot t} = \frac{\dot{m}_{Prop}}{m_{Cat}} \qquad \left[ \frac{kg_{PP}}{g_{Cat} \cdot h} \right] \qquad (4.5)$$

Plotting the current activity over reaction time gives the kinetic profile of the polymerization reaction. Figure 18 shows the derived activity profile of a polymerization carried out without a prepolymerization step. Here, the catalyst was directly injected under reaction conditions. In the first minutes, the system needed time to stabilize (grey area). After approx. 10 to 12 min, constant reaction temperature and pressure were reached and the calculated activity is equal to the real catalyst activity in the reactor. The fluctuations of the activity profile are due to control oscillations of the propylene MFC and are not related to the polymerization reaction itself.

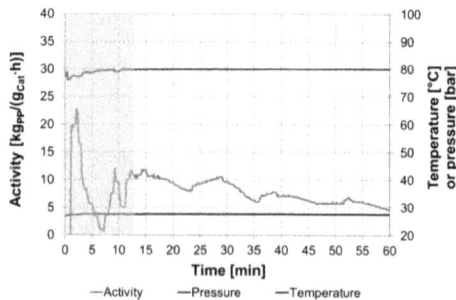

Figure 18: Gained experimental data of the gas-phase polymerization: Activity profile of catalyst B - without prepolymerization

Figure 19 shows the derived activity profile for a polymerization carried out with a prepolymerization step. In the first 15 min, where the prepolymerization occurs, the reactor is heated and filled up with propylene at maximum flow rate. During this time kinetic information are not accessible. Approximately 20 to 25 min after the catalyst injection, constant reaction conditions were reached and the current activity of the catalyst can be calculated directly from the measured mass flow of consumed propylene according equation (4.5).

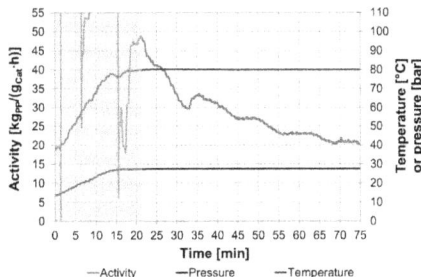

Figure 19: Gained experimental data of the gas-phase polymerization: Activity profile of catalyst B - with prepolymerization

In order to compare the two different ZN catalysts and to describe the influence of the different reaction conditions on the polymerization results, an average activity was used. It is defined as the produced amount of polymer per amount catalyst in one hour (equation 4.6). Experimentally, the average activity was derived from the weighted amount polypropylene per amount catalyst (yield) related to one hour of reaction.

$$\bar{A} = \frac{m_{PP}}{m_{Cat} \cdot h} \qquad \left[ \frac{kg_{PP}}{g_{Cat} \cdot h} \right] \qquad (4.6)$$

### 4.6.2 Test of reproducibility

When performing kinetic measurements, it is crucial to test the reproducibility of the polymerization experiment. Therefore, several experiments were carried out at same reaction conditions. Figure 20 shows the activity profiles of three polymerization reactions carried out without prepolymerization at 80°C with catalyst A. It can be seen that after the initial phase, when constant reaction conditions where reached, the derived activity profiles of all three experiments are comparable. The corresponding average activities as well as the measured MFR values are summarized in Table 9. In general, a standard error of about 10% of experimental measurements has to be assumed due to the high sensitivity and small feeding amounts of the used catalysts.

43

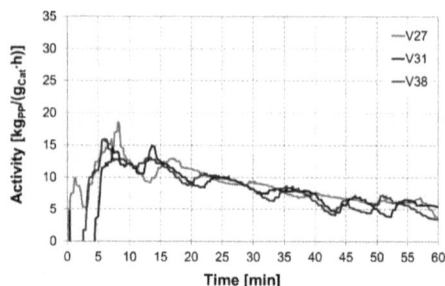

Figure 20: Reproducibility of gas-phase polymerization: Activity profiles (T=80°C, p=27.5 bar, H$_2$=0.025 mol/l, without prepolymerization, catalyst A)

Table 9: Reproducibility of gas-phase polymerization: Average activities and MFR values (T=80°C, p=27.5 bar, H$_2$=0.025 mol/l, without prepolymerization, catalyst A)

| Exp. | $m_{Cat}$ | H$_2$ | Avg. activity | MFR |
|------|------|------|------|------|
| Unit | mg | mol/l | kg$_{PP}$/(g$_{Cat}$·h) | g/10 min |
| V27 | 19.0 | 0.240 | 9.0 | 75.2 |
| V31 | 17.7 | 0.245 | 8.7 | 70.8 |
| V38 | 16.8 | 0.250 | 8.5 | 79.4 |

### 4.6.3 Equilibrium monomer concentration in the polymer particle

For the discussion of the experimental results, the monomer concentration in the polymer particle at the active sites needs to be considered. In gas-phase polymerization, the equilibrium monomer concentration ($c_M$*) can be calculated using Henry's law and the equation of Stern (see chapter 5.1.8 equation (5.52) and (5.53)). Table 10 shows the resulting equilibrium monomer concentrations in the amorphous fraction of the polymer particle at the different investigated reaction conditions.

Table 10: Equilibrium monomer concentration in the polymer particle according Henry's law and Stern equation

| T | p | k* | $c_M$* | $c_M^{GP}$ |
|------|------|------|------|------|
| °C | bar | mol/(atm·l$_{amorph}$) | mol/l | mol/l |
| 50 | 16.4 | 0.0999 | 1.616 | 0.779 |
| 60 | 19.8 | 0.0828 | 1.617 | 0.943 |
| 70 | 25 | 0.0697 | 1.720 | 1.234 |
| 70 | 27.5 | 0.0697 | 1.892 | 1.447 |
| 70 | 30 | 0.0697 | 2.064 | 1.715 |
| 80 | 25 | 0.0596 | 1.470 | 1.138 |
| 80 | 27.5 | 0.0596 | 1.616 | 1.314 |
| 80 | 30 | 0.0596 | 1.763 | 1.515 |
| 90 | 27.5 | 0.0515 | 1.399 | 1.216 |

$c_M$* equilibrium monomer concentration in the amorphous fraction of semi-crystalline polymer

$c_M^{GP}$ monomer concentration of gas-phase in reactor according NIST database [132]

44

Following conclusions can be made: At constant reaction temperature, the equilibrium monomer concentration in the polymer particle $c_M^*$ increases with increasing reaction pressure. In contrast, at constant reaction pressure, $c_M^*$ decreases with increasing reaction temperature. Comparing $c_M^*$ with the monomer concentration of the surrounding gas-phase ($c_M^{GP}$) it can be seen that $c_M^*$ is higher than $c_M^{GP}$. This was already confirmed by Hutchinson and Ray [99] who investigated monomer sorption effects into semi-crystalline polymers.

## 4.6.4 Results catalyst A - Influence of the reaction conditions on catalyst activity

In the following, the influence of the different reaction conditions on the average activity, the activity profile as well as weight average molecular weight are shown. The influences of pressure, temperature and $H_2$ concentration were investigated for polymerizations carried out without a prepolymerization step according procedure 1 (chapter 4.3.2). Herein, the catalyst was directly injected at reaction conditions. With procedure 2, where polymerizations were carried out with a prepolymerization step (chapter 4.3.3), the influence of the catalyst injection conditions will be clarified. The overview of all experiments with detailed reaction conditions and results are summarized in Table 28 to Table 30 in the appendix 9.1.1.

### 4.6.4.1 Influence of reaction pressure

Figure 21 shows the resulting average (avg.) activities for polymerizations carried out at monomer pressures of 25, 27.5 and 30 bar at 80 °C and with a $H_2$ concentration of 0.025 mol/l. It can be seen that the avg. activity linearly increases with increasing reaction pressure in the investigated pressure range. In general, a higher monomer pressure in the gas-phase leads to a higher equilibrium monomer concentration in the polymer particle (see chapter 4.6.3). Thus, the monomer concentration at the active sites of the catalyst is higher which leads to an increase of the reaction rate resulting in an increase of the avg. activity.

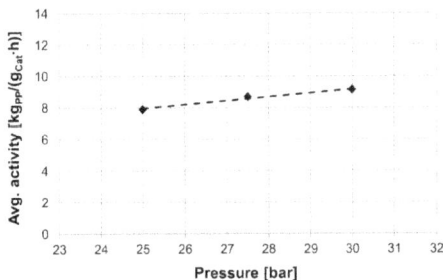

Figure 21: Results gas-phase polymerization catalyst A: Influence of reaction pressure on avg. activity (T=80°C, $H_2$=0.025 mol/l, no prepoly)

Figure 22 a) shows the corresponding activity profiles at the different reaction pressures. As it can be seen, the activity profiles show no significant differences for different reaction pressures. The activity level slightly increases with increasing pressure. When normalizing the activities to the calculated equilibrium monomer concentration $c_M^*$ (Figure 22 b), it can be seen that the activity profiles are quite similar. In the investigated pressure range, the pressure and the monomer concentration, respectively, have therefore no effect on the activity profiles itself. The strong change of the activity for the polymerization at 30 bar after approx. 37 min is due to changes in temperature control mode and has no kinetic background.

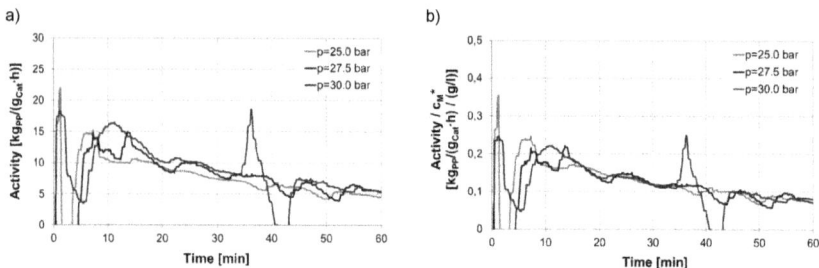

Figure 22: Results gas-phase polymerization catalyst A: Influence of reaction pressure on activity profiles: a) activity, b) activity normalized with equilibrium monomer concentration (T=80°C, H$_2$=0.025 mol/l, no prepoly)

### 4.6.4.2 Influence of reaction temperature

In order to investigate the influence of the reaction temperature on the catalyst activity, polymerization experiments were carried out at temperatures between 50°C to 90 °C with a H$_2$ concentration of 0.025 mol/l (further reaction conditions see chapter 4.4).

Figure 23 shows the resulting avg. activities over the investigated temperature range. As it can be seen, the course of activity as a function of reaction temperature is divided into two parts: From 50°C to 70°C the avg. activity increases with increasing reaction temperature, which is a normal dependency between activity and temperature. With increasing reaction temperature above 70°C, the avg. activity strongly decreases, which would principally not be expected.

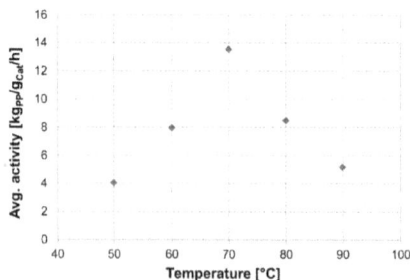

Figure 23: Results gas-phase polymerization catalyst A: Influence of reaction temperature on avg. activity (H$_2$=0.025 mol/l, no prepoly)

46

Figure 24 shows the corresponding activity profiles of catalyst A for the investigated reaction temperatures from 50°C to 90°C. In general, the activity increases rapidly after the reaction start reaching a maximum. Then, the activity decreases over the polymerization time due to the deactivation of the catalyst. The decrease of activity is stronger in the early reaction phase due to the higher activity. At the end only a slight decrease is visible. From 50°C to 70°C the general activity level increases. From 70°C to 90°C the activity level decreases again. This is in accordance with the derived avg. activities shown above. As described before, the current activity at the reaction start could not be determined because the system needed time to stabilize after the catalyst was injected (grey area).

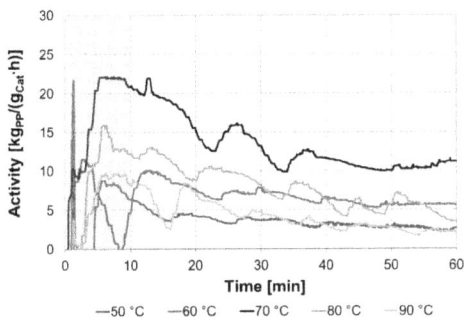

Figure 24: Results gas-phase polymerization catalyst A: Influence of reaction temperature on activity profiles (H$_2$=0.025 mol/l, no prepoly)

As described in chapter 4.6.3, the equilibrium monomer concentration $c_M^*$ decreases with increasing reaction temperature at constant reaction pressure. In order to clarify if the monomer concentration influences the temperature behavior, the average activity normalized with the equilibrium monomer concentration is compared with the average activity in Figure 25.

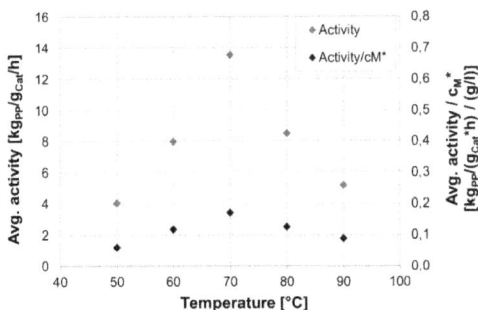

Figure 25: Results gas-phase polymerization catalyst A: Avg. activity normalized with equilibrium monomer concentration (H$_2$=0.025 mol/l, no prepoly)

The avg. activities normalized with $c_M^*$ are showing the same temperature behavior as the avg. activities. The normalized avg. activities increase with increasing reaction temperature up to 70°C where a maximum is reached. With further temperature increase above 70°C, a clear decrease of the values is visible. That means that monomer sorption into the polymer particle cannot be the only explanation of the unusual temperature behavior of the avg. activity with increasing reaction temperature.

A further possible explanation for the decrease of activity with increasing temperatures above 70°C could be a stronger spontaneous deactivation of the active sites over the reaction time. In order to compare the deactivation profiles at different temperatures, the activity profiles shown in Figure 24 were normalized with the corresponding avg. activity and are shown in Figure 26.

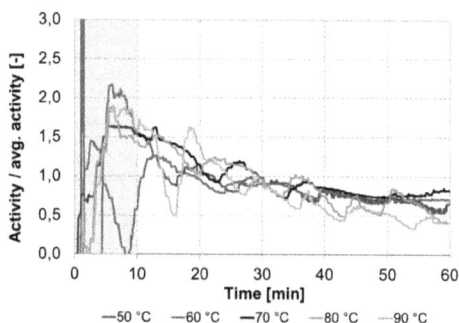

Figure 26: Results gas-phase polymerization catalyst A: Activity profiles normalized with corresponding avg. activities ($H_2$=0.025 mol/l, no prepoly)

It can be seen that the deactivation behavior is similar for the different reaction temperatures from 50°C to 80°C. That means that there is no significant influence of the temperature on the spontaneous deactivation of active sites over the reaction time. The activity profile at 90 °C shows a slightly stronger deactivation at the end of the reaction. Nevertheless, a spontaneous catalyst deactivation could not sufficiently explain the strong decrease in activity at this high reaction temperature.

### 4.6.4.3 Influence of catalyst injection conditions – effect of prepolymerization

In order to examine the start-up behavior of the catalyst, polymerizations were carried out with a prepolymerization step (procedure 2, see chapter 4.3.3). Herein, the catalyst and cocatalyst were injected at milder reaction conditions (40°C and approx. 13.5 bar). The main polymerization was carried out in the temperature range from 70°C to 90 °C at 27.5 bar and with a $H_2$ concentration of 0.025 mol/l. In Figure 27, the resulting avg. activities of these experiments are shown together with the avg. activities of the polymerizations carried out without prepolymerization.

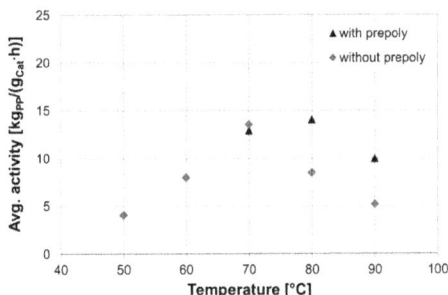

Figure 27: Results gas-phase polymerization catalyst A: Influence of catalyst injection conditions on avg. activity - Comparison polymerization with and without prepolymerization step ($H_2$=0.025 mol/l)

At 70°C, nearly the same avg. activities were reached with both procedures. For high reaction temperatures of 80°C and 90 °C higher avg. activities were reached for polymerizations with prepolymerization step than for polymerizations without prepolymerization.

Figure 28 shows the corresponding activity profiles for reactions at 70°C and 90°C for both procedures during main polymerization (without prepolymerization stage, $t_0=t_{poly}$).

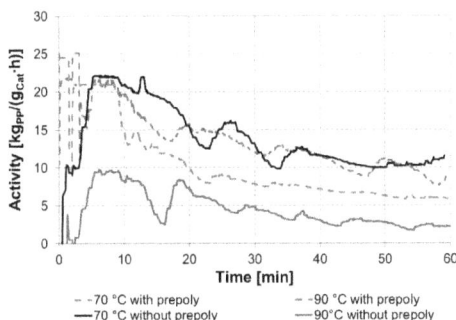

Figure 28: Results gas-phase polymerization catalyst A: Influence of catalyst injection conditions on activity profile - Comparison polymerization with and without prepolymerization step (T=70/90°C, $H_2$=0.025 mol/l)

At 70°C, the activity profiles of polymerizations with and without prepolymerization step are quite similar. In contrast, at 90 °C, the influence of the prepolymerization is much more pronounced. With preceding prepolymerization step, the activity is approx. the double compared to the activity reached with direct injection at 90°C without prepolymerization.

Figure 29 shows the same activity profiles normalized with the corresponding avg. activity. It can be seen that, independently of reaction temperature and injection conditions, the deactivation behavior of the catalyst over reaction time is similar.

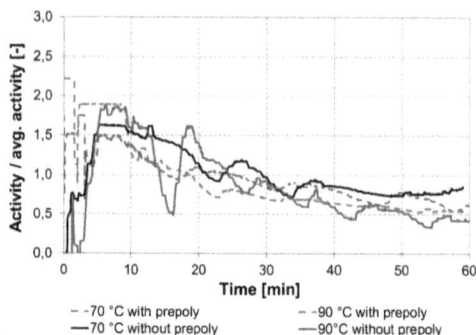

Figure 29: Results gas-phase polymerization catalyst A: Influence of catalyst injection conditions - Activity profiles normalized with corresponding avg. activity (T=70/90°C, $H_2$=0.025 mol/l)

The strong effect of prepolymerization on polymerization activity at 90°C suggests that without prepolymerization the observed decrease in polymerization activity at high temperatures can be indeed - at least partly - related to start-up of the catalyst at the beginning of the reaction due to the harsher conditions.

## 4.6.4.4 Influence of hydrogen concentration

As described in chapter 2.3, hydrogen acts as chain transfer agent and is used to control the molecular weight of the formed polymer. Furthermore, it is also known that it influences the polymerization rate and the activity of the catalyst. In order to study the influence of the hydrogen ($H_2$) concentration on the catalysts activity and the molecular weight of the polymer product, experiments were carried out with different $H_2$ concentrations ranging from 0.0 to 0.05 mol/l at different reaction temperatures ranging from 70°C to 90°C at 27.5 bar. The polymerizations were carried out without a prepolymerization step according procedure 1 (see chapter 4.3.2). It has to be mentioned that for the comparison the given $H_2$ concentration is referred to $H_2$ concentration in the gas-phase calculated from the amount of $H_2$ added to the reactor per reactor volume.

Influence of hydrogen concentration on activity

Figure 30 shows the resulting avg. activities for the investigated $H_2$ concentrations at the three different reaction temperatures.

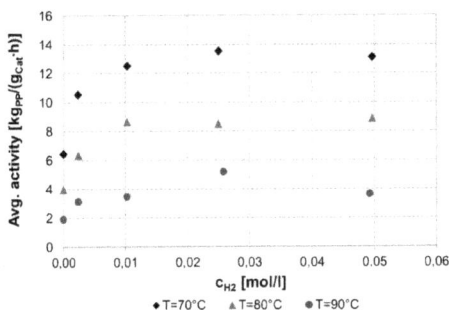

Figure 30: Results gas-phase polymerization catalyst A: Influence of $H_2$ concentration on avg. activity (p=27.5 bar, no prepoly)

At constant reaction temperature, the avg. activity increases with increasing $H_2$ concentration up to 0.025 mol/l. A further increase of the $H_2$ concentration up to 0.05 mol/l does not lead to higher activities. Thus, an activity plateau is reached for $H_2$ concentrations higher than 0.025 mol/l. The trend is visible over the whole investigated temperature range.

During the performance of the experiments it was observed that the $H_2$ concentration has a slight influence on the start of the reaction. Figure 31 shows the activity profiles together with the course of the reaction temperature of polymerizations carried out at 80°C and 27.5 bar without and with hydrogen up to 0.05 mol/l.

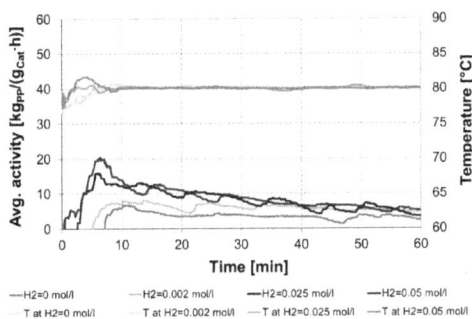

Figure 31: Results gas-phase polymerization catalyst A: Influence of hydrogen on activity profile and reaction temperature (T=80°C, p=27.5 bar, no prepoly)

Two effects can be observed during the reaction start: Firstly, from the activity profile it can be seen that an increasing $H_2$ concentration leads to a slightly faster reaction start. The polymerization with 0.05 mol/l $H_2$ is directly started after the injection of the catalyst/cocatalyst, while the polymerizations without $H_2$ and with 0.002 mol/l $H_2$ is started with a certain delay. Also the activity maximum reached after the reaction start is higher with higher $H_2$ concentrations. (Notice, absolute values of the activity cannot be taken due to the not constant reaction conditions.) Secondly, the faster reaction start for higher $H_2$ concentrations is visible in the temperature profile. With increasing $H_2$ concentration the temperature increase at the beginning of the reaction is also stronger. High concentrations lead to a fast temperature increase and a slight temperature overshoot caused by the released heat from the exothermal polymerization reaction. Without $H_2$ and at low $H_2$ concentration of 0.002 mol/l the temperature increases slowly and does not show a temperature overshoot.

Figure 32 shows the activity profiles normalized with the corresponding avg. activity. It can be seen that for the different $H_2$ concentrations the deactivation profiles are similar. Therefore, $H_2$ has no significant influence on the deactivation behavior of the catalyst in the investigated $H_2$ range.

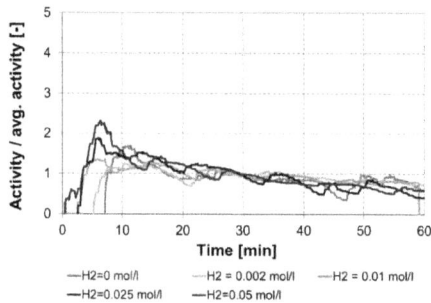

Figure 32: Results gas-phase polymerization catalyst A: Influence of hydrogen on activity - Activity profiles normalized with corresponding avg. activity (T=80°C, p=27.5 bar, no prepoly)

Influence of hydrogen concentration on molecular weight

The weight average molecular weight ($M_w$) of the produced polymer powders were estimated from MFR measurements as described in chapter 4.5.1. Figure 33 shows the resulting $M_w$ of the polymer powder in dependence of the different $H_2$ concentrations for the different investigated reaction temperatures.

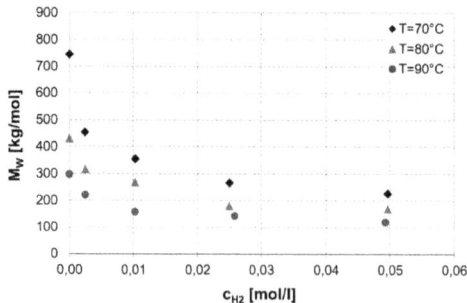

Figure 33: Results gas-phase polymerization catalyst A: Influence of $H_2$ concentration on weight average molecular weight (p=27.5 bar, no prepoly)

As expected, the $M_W$ decreases with increasing $H_2$ concentration at all reaction temperatures. The highest $M_W$ is reached without $H_2$. At higher $H_2$ concentrations of 0.025 and 0.05 mol/l, the reached $M_W$ are very similar. Comparing the $M_W$ at same $H_2$ concentrations but different reaction temperatures, it can be seen that always lower $M_W$ were reached with increasing temperature from 70°C to 90°C. The differences are more pronounced for reactions without $H_2$ and with lower $H_2$ concentrations.

### 4.6.5 Results catalyst B - Comparison of Ziegler-Natta catalysts of different activity

Kinetic measurements of the gas-phase polymerization of propylene were carried out with a second ZN catalyst B. In general, catalyst B has a higher activity compared to catalyst A. The kinetic measurements were carried out within the same operating window as for catalyst A: The studied temperature range was between 50 °C and 90 °C, the reaction pressures between 25 and 30 bar and the hydrogen concentrations between 0 and 0.05 mol/l. The influence of the different reaction conditions on catalyst activity as well as on $M_W$ were investigated for polymerizations carried out without a prepolymerization step. Also polymerization experiments with a preceding prepolymerization step were carried out in order to study the influence of the catalyst injection conditions on the reachable activities.

In the following, the resulting avg. activities as well as $M_w$ will be shown for catalyst B in comparison with catalyst A. The results as well as the reaction conditions are summarized in detail in Table 28 to Table 30 in appendix 9.1.1. Selected kinetic profiles are also shown as similar dependencies were derived for catalyst B as already discussed for catalyst A.

#### 4.6.5.1 Influence of reaction pressure

Figure 34 shows average activities of catalyst A and B at the different reaction pressures from 25 to 27.5 bar at 80°C for catalyst A and at 70°C for catalyst B. In all experiments the $H_2$ concentration was 0.025 mol/l.

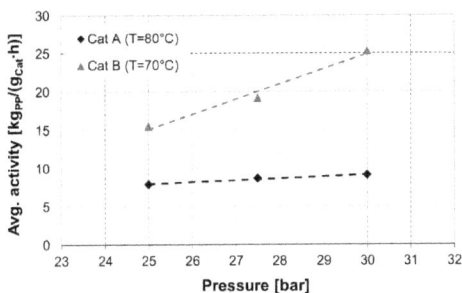

Figure 34: Results gas-phase polymerization – Comparison catalyst A and B: Influence of reaction pressure on avg. activity (cat A T=80°C, cat B T=70°C, H₂=0.025 mol/l, no prepoly)

As expected, the average activity of catalyst B also linearly increases with increasing reaction pressure. As explained before, an increasing pressure leads to an increase of the monomer concentration at the active sites resulting in higher avg. activity, see Table 11. Furthermore, the increase of the avg. activity of catalyst B is stronger compared to catalyst A with increasing reaction pressure. The stronger pressure effect for catalyst B compared to catalyst A can be explained by the generally higher activity of catalyst B. Table 11 shows the corresponding equilibrium monomer concentration $c_M^*$ at 70°C and 80°C. Even though $c_M^*$ is slightly higher at 70°C as at 80°C it can be seen that, when relating the avg. activity to the monomer concentration, the results for catalyst B are on a higher level compared to catalyst A.

Table 11: Results gas-phase polymerization – Comparison catalyst A and B: Influence of $c_M^*$ on avg. activity ($H_2$=0.25 mol/l, no prepoly)

| Exp. | catalyst | T | p | $c_M^*$ | A | $A/c_M^*$ |
|------|----------|-----|-----|---------|----------------------------|--------------------------------------|
|      |          | °C  | Bar | mol/l   | $kg_{PP}/(g_{Cat} \cdot h)$ | $(kg_{PP}/(g_{Cat} \cdot h)/(g/l)$ |
| V30 | A | 80 | 25 | 1.470 | 7.88 | 0.127 |
| V31 | A | 80 | 27.5 | 1.617 | 8.68 | 0.127 |
| V29 | A | 80 | 30 | 1.763 | 9.11 | 0.122 |
| V21 | B | 70 | 25 | 1.720 | 15.49 | 0.214 |
| V7 | B | 70 | 27.5 | 1.892 | 19.18 | 0.241 |
| V22 | B | 70 | 30 | 2.064 | 25.30 | 0.291 |

$^*c_M$: equilibrium monomer concentration in polymer particle calculated according Henry's law and Stern equation

## 4.6.5.2 Influence of reaction temperature

Figure 35 shows the resulting avg. activities of both catalysts over the investigated temperature range from 50°C to 90°C at 27.5 bar and with 0.025 mol/l $H_2$.

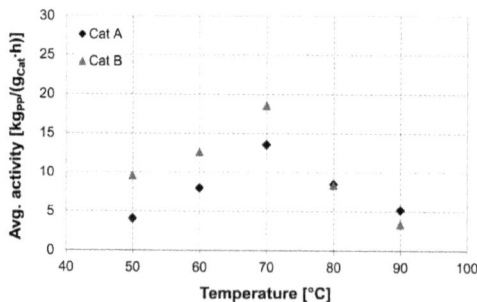

Figure 35: Results gas-phase polymerization – Comparison catalyst A and B: Influence of reaction temperature on avg. activity ($H_2$=0.025 mol/l, no prepoly)

The course of activity of catalyst B is also divided into two parts as for catalyst A. From 50 °C to 70°C the avg. activity increases with increasing reaction temperature and strongly decreases with further increase of reaction temperature above 70°C. Comparing both catalysts it can be seen that from 50°C to 70°C always higher activities were reached with catalyst B. This can be explained by its generally higher activity compared to catalyst A. A further increase of the reaction temperature above 70°C leads to a stronger decrease of the avg. activity of catalyst B compared to catalyst A. The effect of higher reaction temperatures on the catalyst activity is therefore stronger for catalyst B than for catalyst A.

The corresponding activity profiles of catalyst B derived at the different reaction temperatures are shown in Figure 36 a). The same trends as described for the avg. activities are obtained. Figure 36 b) shows the activity profile normalized with the corresponding avg. activity in order to compare the deactivation behavior of the catalyst at the different reaction temperatures. It can be seen that the deactivation behavior is similar for the different reaction temperatures from 50°C to 80°C. The normalized activity profile at 90°C shows a slightly stronger deactivation behavior.

a)

b)

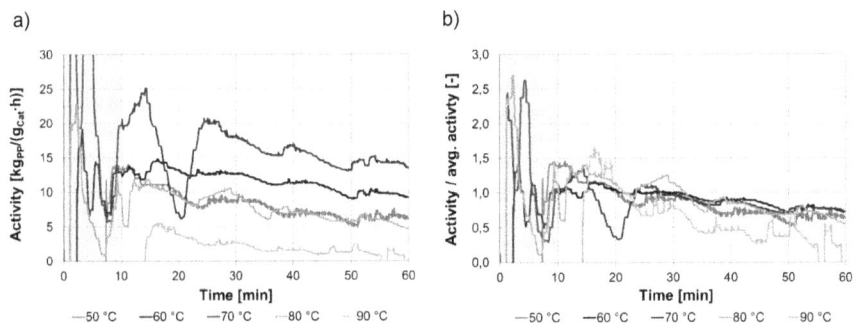

Figure 36: Results gas-phase polymerization catalyst B: a) Influence of reaction temperature on activity profile, b) Activity profile normalized with corresponding avg. activity ($H_2$=0.025 mol/l, no prepoly)

### 4.6.5.3 Influence of catalyst injection conditions – effect of prepolymerization

The effect of prepolymerization on the catalyst activity was also studied with catalyst B in the temperature range from 60°C to 90°C and with 0.025 mol/l $H_2$. The resulting avg. activities are shown in comparison with the avg. activities derived without prepolymerization in Figure 37.

For catalyst B, the effect of prepolymerization is already visible at 60°C. At higher reaction temperatures, much higher activities (up to three times) were reached when a prepolymerization was applied.

55

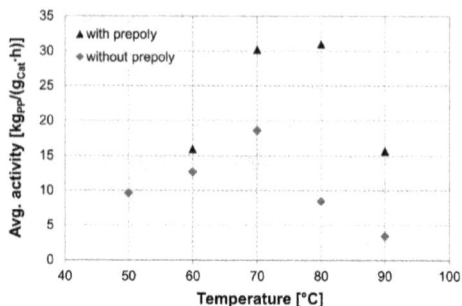

Figure 37: Results gas-phase polymerization catalyst B: Influence of catalyst injection conditions on avg. activity ($H_2$=0.025 mol/l)

The derived activity profiles at 60°C and 80°C are shown for both procedures in Figure 38 a). When normalizing the activity profiles with the corresponding avg. activity (Figure 38 b) it can also be seen for catalyst B that the injection conditions have no influence of the course of activity itself.

a)

b)

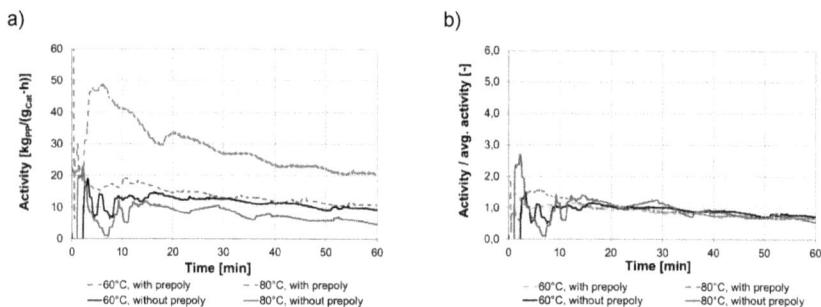

Figure 38: Results gas-phase polymerization catalyst B: a) Influence of catalyst injection conditions on activity profile, b) Activity profiles normalized with avg. activity ($H_2$=0.025 mol/l)

In Figure 39, the influence of the different catalyst injection conditions on the resulting activities is shown for both ZN catalysts. In order to compare the different catalysts, the avg. activities derived without prepolymerization were normalized with the avg. activities reached with prepolymerization.

56

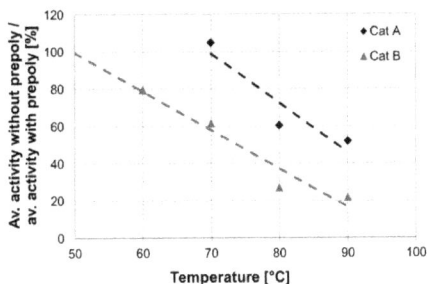

Figure 39: Results gas-phase polymerization – Comparison catalyst A and B: Influence of catalyst injection conditions – avg. activity without prepolymerization normalized with avg. activity with prepolymerization (activity cat A < activity cat B)

For both ZN catalysts, the avg. activities strongly decrease with increasing reaction temperature when no prepolymerization step was applied. For the higher active catalyst B, the influence of the prepolymerization on the activity is already visible at 60 °C, whereas for the less active catalyst A the influence of the prepolymerization is visible only at higher reaction temperatures above 70°C. Furthermore, it can be seen that for catalyst B the influence of the prepolymerization step at higher temperatures is more pronounced than for catalyst A. For example, for catalyst B at 90°C only 20 % of the activity is reached for polymerization without prepolymerization compared to polymerization with prepolymerization, whereas for catalyst A approx. 50 % of the activity is reached.

### 4.6.5.4 Influence of hydrogen concentration

The influence of hydrogen on the catalyst activity as well as on the molecular weight was also studied for catalyst B. Therein, polymerizations (without prepolymerization step) were carried out without $H_2$ and with $H_2$ concentrations from 0.002 to 0.05 mol/l within a temperature range from 70°C to 90°C. Figure 40 and Figure 41 show the derived avg. activities and weight average molecular weights of the resulting polymer as a function of the different $H_2$ concentrations.

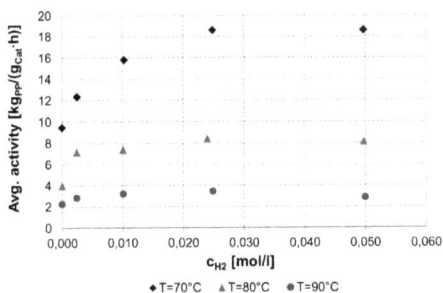

Figure 40: Results gas-phase polymerization catalyst B: Influence of $H_2$ concentration on avg. activity (p=27.5 bar, no prepoly)

57

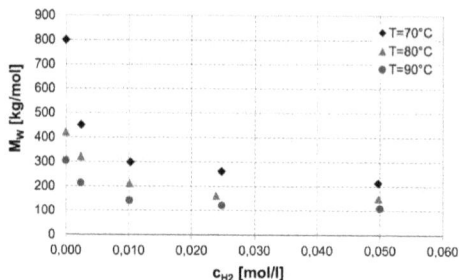

Figure 41: Results gas-phase polymerization catalyst B: Influence of $H_2$ concentration on weight average molecular weight (p=27.5 bar, no prepoly)

Same tendencies of the hydrogen response as described for catalyst A can also be seen for catalyst B. With increasing $H_2$ concentration up to 0.025 mol/l the avg. activity increases. A further increase of $H_2$ concentration up to 0.05 mol/l does not lead to a further activity increase. The molecular weights decrease with increasing $H_2$ concentration as expected. Furthermore, higher molecular weights are derived at lower reaction temperatures.

Figure 42 shows a direct comparison of the hydrogen response in respect to the avg. activity for both catalysts. Herein, the avg. activities at each $H_2$ concentration were normalized with the corresponding avg. activity reached at the highest $H_2$ concentration of 0.05 mol/l for a reaction temperature of 80°C. It can be seen that catalyst A and B have nearly the same hydrogen response within the investigated $H_2$ range.

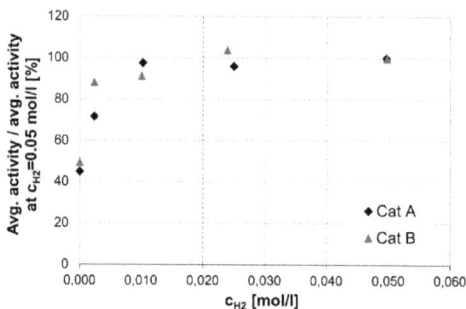

Figure 42: Results gas-phase polymerization – Comparison hydrogen response catalyst A and B: Avg. activity normalized with avg. activity at $c_{H2}$=0.05 mol/l (T=80°C, p=27.5 bar, no prepoly))

### 4.6.6 Results polymer characterization

The produced polymer powders were analyzed using different analytical methods described in chapter 4.5. In the following, results of selected polymer samples are presented showing the influence of the different reaction conditions on the polymer properties. A detailed overview of the results is given in appendix 9.1.2.

### 4.6.6.1 Crystallinity

The crystallinity of the polymer samples was determined using differential scanning calorimetry (chapter 4.5.2). The crystallinity was calculated from the specific heat enthalpy of the first melting peak of the heat flow curve related to the specific enthalpy of a 100 % crystalline polymer according equation (4.2).

The crystallinity of polypropylene samples produced with catalyst A at 70°C and 90°C (both with 0.025 mol/l $H_2$, without prepolymerization) was measured directly after the polymerization. The resulting crystallinity of the polypropylene powder sample produced at 70°C was 35.6 % (±1.1 %) and at 90°C was 38.5 % (±1.6 %).

Furthermore, polymer powders produced at different reaction temperatures and $H_2$ concentrations as well as different injection conditions were analyzed for both catalysts. The resulting crystallinities as well as the corresponding melting temperatures are summarized in Table 12. It has to be mentioned that these measurements were carried out at later times without stabilizing the polymer samples. The resulting crystallinities are therefore higher as the crystallinities measured directly after the reaction due to post-crystallization.

Table 12: Results analytics – DSC measurements: Crystallinity of polymer produced with catalyst A and B at different reaction conditions

| T | $H_2$ | Prepoly | Crystallinity Cat A | Crystallinity Cat B |
|---|---|---|---|---|
| °C | mol/l | | % | % |
| 50 | 0.025 | no | - | 54.3 |
| 70 | 0 | no | 50.2 | 49.5 |
| 70 | 0.025 | no | 52.3 | 51.3 |
| 70 | 0.025 | yes | 52.5 | 51.1 |
| 70 | 0.05 | no | 54.3 | 51.1 |
| 90 | 0.025 | no | 49.2 | 49.5 |
| 90 | 0.025 | yes | 51.9 | - |

For both catalysts following tendencies could be observed: At constant reaction temperature of 70°C, the crystallinity slightly increases with increasing $H_2$ concentration. At constant $H_2$ concentration, the crystallinity slightly decreases with increasing reaction temperature. Comparing polymerizations carried out with and without prepolymerization step at same reaction conditions, it can be seen that the prepolymerization has no significant effect on the crystallinity.

## 4.6.6.2 Molecular weights and molecular weight distribution

The weight average molecular weights were determined from MFR measurements as described in chapter 4.5.1. The resulting $M_w$ for each polymer sample are summarized in Table 28 to Table 30 in the appendix 9.1.1.

From selected polymer samples produced with catalyst B additional GPC measurements were carried out by the external collaborator in order to confirm the $M_w$ and to determine the broadness of the molecular weight distribution. The resulting weight avg. molecular weights derived from both measurements are summarized in Table 13. Except of polymer produced at 70°C without hydrogen (V9), $M_w$ determined from MFR measurements are in a good agreement with $M_w$ derived from GPC measurements. The deviation of the samples V9 is due to the relatively high molecular weight compared to the other samples and can be related to the GPC integration method. The average polydispersity index (PDI), which was calculated from GPC measurements by relating $M_w$ to $M_n$, is 6. This higher PDI is typically for polymer produced with ZN catalysts due to the multiple active sites resulting in a broader MWD (see chapter 2.1.1).

Table 13: Results analytics – Comparison weight avg. molecular weights derived from MFR and GPC measurements – reaction conditions (polymer produced with catalyst B)

| Exp. | Conditions | | | MFR | GPC | | |
|---|---|---|---|---|---|---|---|
| | T | $H_2$ | Prepoly | $M_w$ | $M_n$ | $M_w$ | PDI |
| No. | °C | mol/l | | kg/mol | kg/mol | kg/mol | - |
| 8 | 50 | 0.025 | no | 230.0 | 34.2 | 211.5 | 6.2 |
| 9 | 70 | 0 | no | 799.9 | 106.2 | 998.1 | 9.4 |
| 13 | 70 | 0.01 | no | 298.6 | 50.1 | 307.7 | 6.2 |
| 27 | 70 | 0.025 | no | 261.0 | 42.9 | 241.4 | 5.6 |
| 23 | 70 | 0.025 | yes | 256.7 | 43.0 | 252.3 | 5.9 |
| 16 | 70 | 0.05 | no | 213.4 | 35.4 | 213.8 | 6.0 |
| 4 | 90 | 0.025 | no | 120.5 | 21.8 | 169.6 | 7.6 |

## 4.6.6.3 Scanning electron microscopy

The particle morphology was analyzed using scanning electron microscopy (see chapter 4.5.6). Aim was to image the influence of the different reaction conditions (reaction temperature, $H_2$ concentration as well as the different catalyst injection conditions) on the morphology of the polymer particles. Furthermore, polymer particles produced with both investigated ZN catalysts are also compared.

Influence of reaction temperature

Figure 43 shows SEM images of polymer powders produced with catalyst B without prepolymerization at different reaction temperatures. At low reaction temperature of 50°C, spherical polymer particles were produced having a smooth surface with small cracks. With increasing reaction temperature, particles show a not perfectly spherical shape with a rough particle surface. At highest reaction temperature of 90°C, non-spherical polymer particles up to polymer flakes were formed. This suggests that the higher the catalyst injection temperature the more the influence on the catalyst break-up and therefore on the final morphology of the polymer particle.

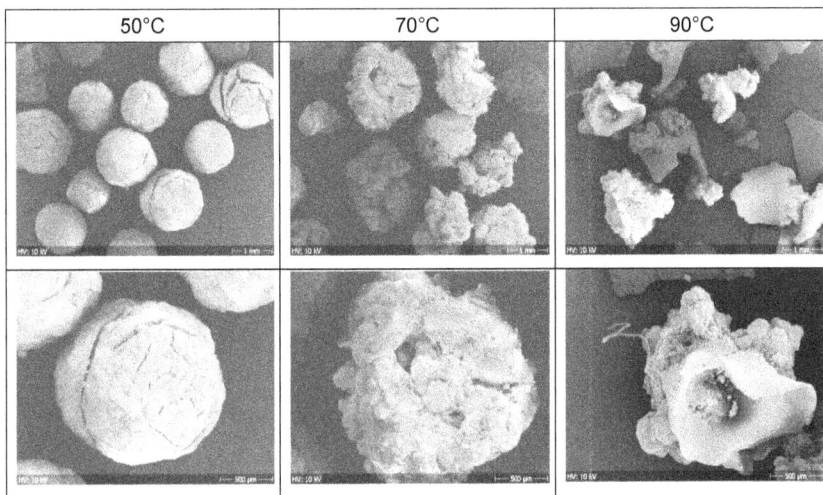

Figure 43: Results analytics – SEM: Influence reaction temperature on polymer morphology (catalyst B, without prepoly)

Influence of injection conditions

Figure 44 shows the comparison of polymer particles produced with and without a prepolymerization step. Furthermore, polymer particles produced only during the prepolymerization are shown. Applying a prepolymerization step means that the catalyst is always injected at lower reaction temperature (40°C) and the reaction conditions are slowly adjusted during prepolymerization. From the SEM images it can be seen that the prepolymer as well as the polymer produced by polymerization with prepolymerization the particles show a similar spherical shape with a smooth surface and small cracks. Therein, the prepolymer replicates the spherical structure of the catalyst particle. With further polymerization, this prepolymer will grow regularly forming larger particles of the same shape. In contrast, and as already shown before, when the catalyst was directly injected at reaction conditions, the polymer particles have an irregular, non-spherical shape with a rough surface. Flat parts on the particle surface suggest that some parts of the particle were spalled.

The comparison clearly shows that the prepolymerization step leads to a better morphology of the polymer particles. A controlled catalyst break-up can be realized and the polymer particle grows regularly.

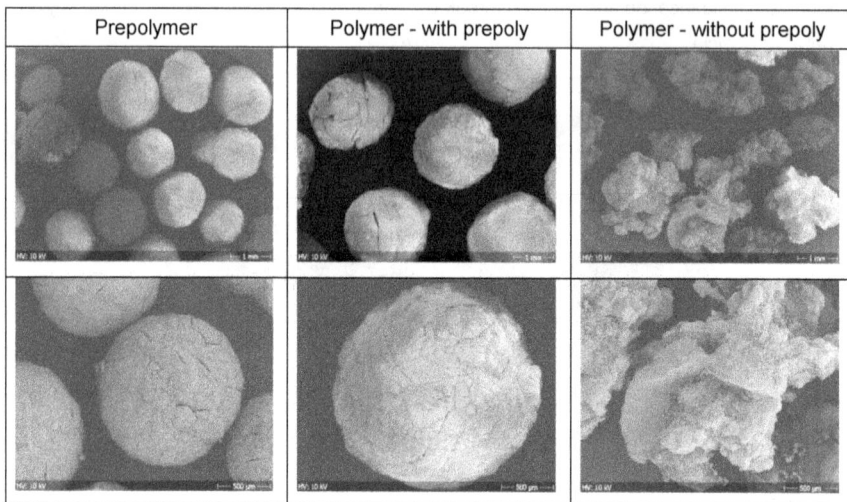

| Prepolymer | Polymer - with prepoly | Polymer - without prepoly |
|---|---|---|
|  |  |  |
|  |  |  |

**Figure 44: Results analytics – SEM: Influence injection conditions on polymer morphology (cat B, T=80°C)**

Comparison particle morphology of both ZN catalysts

Figure 45 shows the comparison of polymer particles produced with both ZN catalysts. In case of polymerization with prepolymerization step, polymer particles produced with catalyst A are also of a spherical shape and have a smooth surface. The particles seem to be smaller and the particle surface show no small cracks compared to polymer particles produced with catalyst B. Without prepolymerization, polymer particles produced with catalyst A show also an irregular shape with a rough surface like for catalyst B (see Figure 44).

Influence of $H_2$ concentration

Figure 46 shows polymer particles produced at 70°C without prepolymerization with different $H_2$ concentrations. From the SEM images no clear influence of the $H_2$ on the particle morphology is visible. The polymer particles showed a non-spherical shape with a rough surface for all $H_2$ concentrations.

| Cat A – without prepoly | Cat A – with prepoly | Cat B – with prepoly |
|---|---|---|

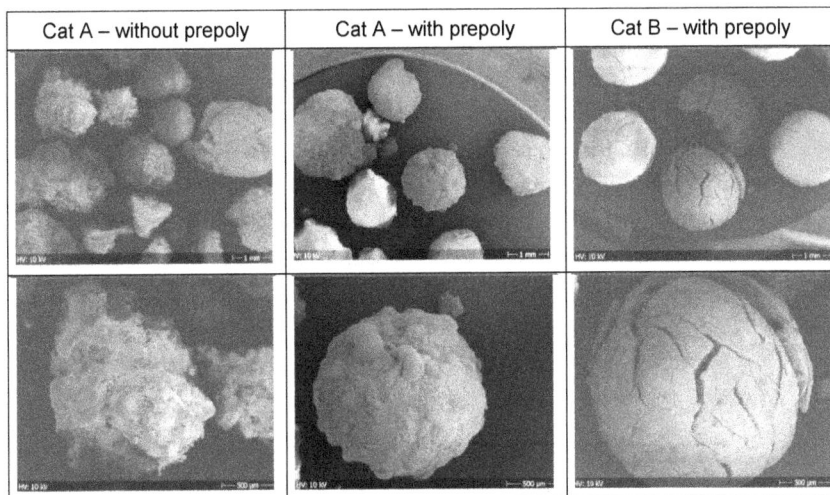

Figure 45: Results analytics – SEM: Comparison polymer morphology catalyst A and B (T=70°C)

| No hydrogen | $H_2$ = 0.01 mol/l | $H_2$ = 0.05 mol/l |
|---|---|---|

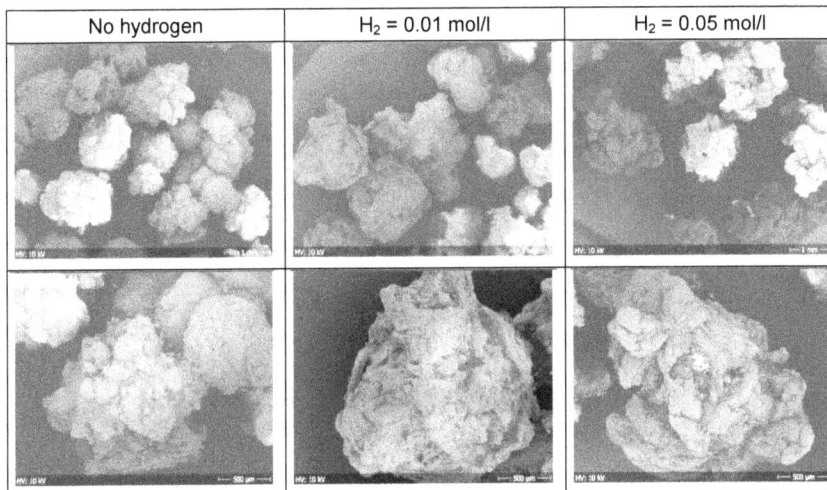

Figure 46: Results analytics – SEM: Influence $H_2$ concentration on polymer morphology (catalyst B, T=70°C, without prepoly)

### 4.6.6.4 Particle size and particle size distribution

For the investigation of the influence of the different reaction conditions on the particle size, polymer samples produced with catalyst A and without prepolymerization were investigated using sieving analysis (see chapter 4.5.5). Therein, the polymer sample was sieved into three main fractions: fines with particle sizes (PS) below 400 µm, middle PS between 400 and 1000 µm and PS above 1000 µm. Furthermore, particle size distribution (PSD) of polymer produced without prepolymerization was measured by the cooperation partner using Malvern analysis. As it was mentioned, the particle fraction above 2000 µm was sieved out before the measurement. A direct comparison between the two measurements is therefore not accurate enough and therefore only trends can be compared. The results of the sieving analysis as well as Malvern analysis are summarized in Table 32 in Appendix 9.1.2.2.

Figure 47 shows the results of the sieving of polymer powders produced at different reaction temperatures. As general tendency it can be seen that with increasing reaction temperature the PS decrease. Fraction above 1000 µm decreases with increasing reaction temperature, while the amount of fines (<400 µm) as well as the fraction between 400 and 1000 µm increase. On the one hand, the strong decrease of activity with increasing reaction temperature above 70°C leads to smaller particle sizes. On the other hand, the rough injection conditions at high temperatures accelerated the formation of fines (see also microscopy pictures). This tendency is also confirmed from Malvern analysis (see Table 32, Appendix 9.1.2.2). With increasing reaction temperatures above 70°C the fractions >2000 µm strongly decrease.

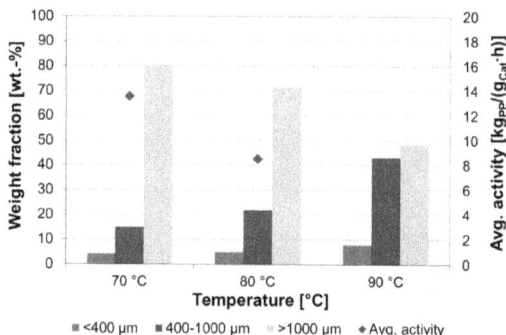

Figure 47: Results analytics – Sieving analysis: Influence of reaction temperature on particle size (catalyst A, $cH_2$=0.025 mol/l, without prepolymerization)

Figure 48 shows the PSD as weight fraction distribution from sieving analysis of polymer samples produced with prepolymerization for both catalysts A and B as well as for polymer produced without prepolymerization for catalyst B under same conditions. The $D_{50}$ values as well as the particle fraction above 2000 µm are summarized in Table 14.

Table 14: Results analytics – Sieving analysis: D50 value and weight fraction above 2000 μm PSD of polymer samples produced with catalyst A and B, with and without prepolymerization (T=70°C, $c_{H2}$=0.025 mol/l)

| Exp. | Cat. | Prepoly | PSD / $D_{50}$ μm | >2000 μm wt.% | Avg. activity $kg_{PP}/(g_{cat} \cdot h)$ |
|------|------|---------|-------------------|---------------|-------------------------------------------|
| V50 | A | yes | 1526 | 23.4 | 12.9 |
| V23 | B | yes | 2117 | 64.7 | 30.2 |
| V27 | B | no | 1773 | 27.0 | 18.6 |

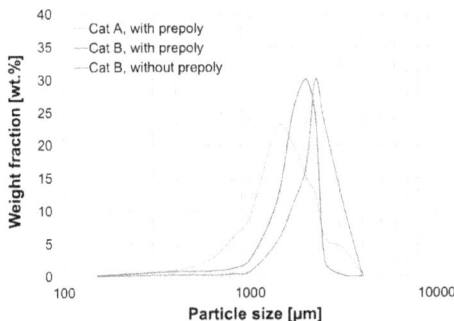

Figure 48: Results analytics – Sieving analysis: PSD of polymer samples produced with catalyst A and B, with and without prepolymerization (T=70°C, $c_{H2}$=0.025 mol/l)

In case of polymerization with prepolymerization, polymer produced with catalyst A shows lower particle sizes (lower $D_{50}$ value) compared to polymer produced with catalyst B. Approx. 23 wt% of the polymer sample produced with catalyst A have particle sizes above 2000 μm, whereas with catalyst B approx. 65 wt% of the polymer sample consists of particles with sizes above 2000 μm. Furthermore, a much broader PSD is derived with catalyst A. As shown in Table 5, the $D_{50}$ value of both catalysts is similar but the activity of catalyst B is higher compared to catalyst A. The much broader PSD of catalyst A might be due to a broader particle size distribution of the catalyst or might also be a hint for break-up of particles and fines formation.

Comparing the PSD of polymer produced with catalyst B but at different injection conditions, it can be seen that without prepolymerization the PSD is shifted to lower particle sizes, whereas the width of the distribution is similar. The $D_{50}$ value of polymer produced without prepolymerization is lower and the particle fraction >2000 μm is much smaller compared to polymerization with prepolymerization. This change to smaller particle sizes can be attributed to both lower activities and break-up of the particles when no prepoly is applied.

### 4.6.6.5 Bulk density

The bulk density was determined gravimetrically as described in chapter 4.5.4. Polymer samples produced with both catalysts at different reaction temperatures, $H_2$ concentrations and injection conditions were analyzed. The results are shown in Figure 49.

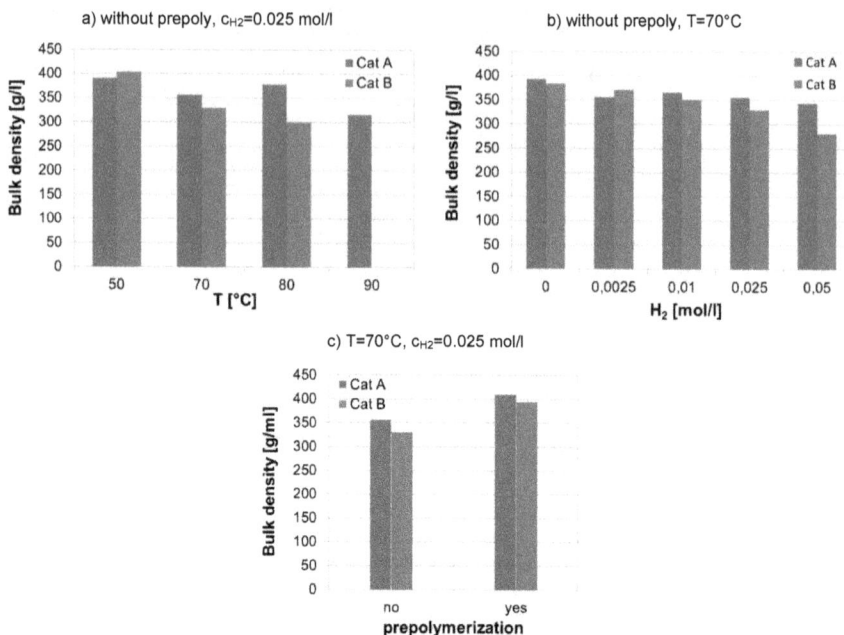

Figure 49: Results analytics – Bulk density: a) Influence of reaction temperature, b) Influence of $H_2$ concentration, c) Influence of injections conditions

In general, slightly higher bulk densities were reached with catalyst A compared to catalyst B. From the comparison of the SEM pictures of the two catalysts (see chapter 4.6.6.3) it could be seen that similar morphologies were reached but with catalyst A smaller particles were produced (see chapter 4.6.6.4, particle size and particle size distribution). Smaller (spherical) particles will lead to a higher bulk density.

It was also shown that the morphology of the polymer particles is mainly influenced by the polymerization temperature as well as the catalyst injection conditions. As shown in Figure 49 a), with increasing reaction temperature the bulk densities decrease. An explanation is that with increasing reaction temperature the particle morphology changes from spherical to non-spherical irregular particles or flakes which influences the bulk densities. This was also examined by Pater et al. [117] and Patzlaff [92] who showed that irregular formed polymer particles have a lower bulk density compared to spherical polymer particles.

From SEM analysis, no significant influence of $H_2$ on the particle morphology could be recognized at 70°C. Comparing the bulk densities in Figure 49 b) it can be seen for both catalysts that with increasing $H_2$ concentration the bulk densities slightly decrease.

Comparing the resulting bulk densities of polymer produced with and without prepolymerization (Figure 49 c), it can be clearly seen that with prepolymerization higher bulk densities were reached. The reason is also the change in particle morphology: With prepolymerization spherical polymer particles were formed, compared to irregular non-spherical particles produced without prepolymerization. This confirms that irregular formed polymer particles lead to lower bulk densities.

## 4.6.6.6 Polymer particle porosity

Porosity of the polymer powders was measured using low and high pressure mercury porosimetry (chapter 4.5.3). The resulting total porosities of polymer samples produced at the different reaction conditions with both catalysts are summarized in Table 15.

Table 15: Results analytics – Porosity measurements: Porosity of polymer powders produced with catalyst A and B at different reaction conditions

| T | $H_2$ | Prepoly | Porosity Cat A | Porosity Cat B |
|---|---|---|---|---|
| °C | mol/l | | % | % |
| 50 | 0.025 | no | | 26.1 |
| 70 | 0 | no | | 23.7 |
| 70 | 0.0025 | no | | 21.4 |
| 70 | 0.025 | no | 26.7 | 24.0 |
| 70 | 0.025 | yes | 23.1 | 23.0 |
| 80 | 0.025 | no | | 24.2 |
| 90 | 0.025 | no | | 30.0 |

From the measurements, no clear trend between reaction temperature as well as $H_2$ concentration and porosity of the polymer samples produced without prepolymerization can be seen. Only for high reaction temperature of 90°C, the polymer porosity is very high compared to other temperatures, which might be due to the different polymer morphology derived at the harsh catalyst injection conditions (see chapter 4.6.6.3). Polymer produced with prepolymerization step shows slightly lower porosities compared to polymer produced without prepolymerization.

In general, for later calculations, an avg. porosity of 25 % for polymer produced with both catalysts and at the different conditions is assumed.

## 4.6.6.7 Sorption measurements

In order to clarify if the different morphology of the polymer particles has an influence on the sorption behavior of the monomer, sorption measurements were carried out with polymer samples produced at lower and higher reaction temperatures as well as under different injection conditions. The measurements were carried out always at 70°C and a monomer pressure of 7 bar (see chapter 4.5.7). In Figure 50, the resulting sorption curves are presented which show the amount of sorbed monomer (propylene) per amount of polymer (polypropylene) over time.

The slope of the mass uptake at the beginning of the measurement gives information about the rate of the sorption, whereas the constant value, which is reached after mass uptake, is equal to the equilibrium gas solubility.

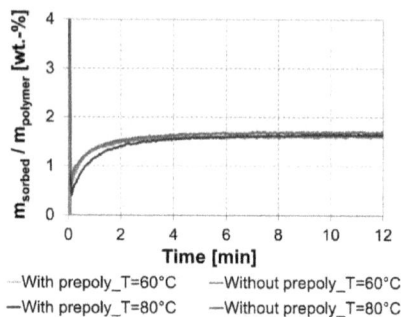

Figure 50: Results analytics – Sorption measurements: Mass uptake curve of propylene in polypropylene particles produced with catalyst B at different reaction conditions (sorption measurements at T=70°C, p=7bar)

Comparing the sorption curves it can be seen that similar equilibrium monomer solubilities as well as comparable mass transfer rates were reached for the polymer samples produced at the different reaction conditions. In contrast, from the SEM images a clearly influence of the reaction temperature as well as the injection conditions on the particle morphology could be seen (chapter 4.6.6.3). It can be therefore concluded that the sorption behavior of the polymer particles is similar for the different particle morphologies derived at the different reaction conditions.

## 4.7 Summary experimental investigation of the gas-phase polymerization of propylene with two different Ziegler-Natta catalyst

Kinetic measurements of the gas-phase polymerization of propylene with two 4th generation Ziegler-Natta catalysts were carried out under industrially relevant conditions. Therein, ZN catalyst A has a lower activity compared to ZN catalyst B. Aim was to investigate the influences of different reaction conditions (reaction temperature, pressure, hydrogen concentration) as well as different catalyst injection conditions (effect of prepolymerization) on the average catalyst activities, the activity profiles as well as on polymer characteristics.

Influence of reaction pressure

For both catalysts, the avg. activity linearly increased with increasing pressure. Due to the pressure increase, the equilibrium monomer concentration at the active sites increases leading to higher reaction rates and therefore to an increase of the avg. activity. The pressure effect is more pronounced for the more active catalyst B compared to catalyst A. An influence of the pressure on the course of the kinetic profile could not be observed.

## Influence of reaction temperature

The comparison of both catalysts showed a similar kinetic behavior: With increasing reaction temperature from 50°C to 70°C the avg. activity increased. A further increase of the reaction temperature above 70°C led to a strong decrease of the avg. activity when no prepolymerization was applied. Furthermore, the comparison of both catalysts showed that from 50°C to 70°C the avg. activity of catalyst B is approx. double as the avg. activity of catalyst A. For higher reaction temperatures above 70°C, the decrease of the avg. activity for the more active catalyst B was stronger compared to catalyst A. The activity profiles of both catalysts derived at the different reaction temperatures showed the same trends as the avg. activities.

Though, the equilibrium monomer concentration in the polymer particle decrease with increasing reaction temperature, it could be shown that monomer sorption into the polymer particle cannot sufficiently explain the strong activity decrease at higher reaction temperatures. Normalizing the activity profiles with the corresponding avg. activity led to a similar deactivation behavior at the different reaction temperatures. Thus, it can be concluded that there is, in general, no significant influence of the higher reaction temperature on the spontaneous deactivation of the catalyst over the reaction time which could explain the strong decrease of the avg. catalyst activity.

## Influence of prepolymerization

The application of a prepolymerization step (catalyst injection at mild conditions) improved the activities of both catalysts at higher reaction temperatures during main polymerization. Therein, the effect of prepolymerization is catalyst specific and depends on the catalyst activity at the reaction temperature. Whereas for catalyst B the effect of the prepolymerization on the avg. activity was already visible at 60°C, higher avg. activities were reached for catalyst A above 70°C. The effect of prepolymerization was more pronounced for the more active catalyst B. Thus, it can be concluded that the higher the catalyst activity is the more critical is the catalyst injection temperature on the final avg. activity.

In contrast, the normalized activity profiles with the corresponding avg. activity of polymerizations with prepolymerization showed no significant differences in the deactivation behavior compared to polymerizations carried out without prepolymerization at the different reaction temperatures. Thus, the catalyst injection conditions have no significant effect on spontaneous catalyst deactivation over time.

The strong effect of prepolymerization at higher reaction temperatures therefore suggests that without prepolymerization the observed decrease in polymerization activity can be - at least partly - related to start-up of the catalyst at the beginning of the reaction due to the harsher catalyst injection conditions. A possible consequence could be an uncontrolled catalyst break-up wherein the polymerization active sites could not be activated and/or the active sites are not accessible during the polymerization. Also local particle overheating due to an insufficient heat transfer area could influence the amount of available active sites for the polymerization reaction (see chapter 2.4). The results therefore suggest that the differences in activity, when a prepolymerization step is applied, might be due to a higher amount of active sites available at the beginning of the reaction.

Influence of hydrogen

Both catalysts showed a similar hydrogen response in respect to the avg. activities as well as molecular weights. As expected, molecular weights decreased with increasing $H_2$ concentration for both catalysts as $H_2$ acts as chain transfer agent. Furthermore, lower molecular weights were derived at higher reaction temperatures. Broad molecular weight distributions with an avg. PDI of 6 were determined from GPC measurements for polymers produced with both ZN catalysts.

For both catalysts, increasing $H_2$ concentrations up to 0.025 mol/l led to an exponential increase of the avg. activities where an activity plateau was reached. A further increase of the $H_2$ concentration up to 0.05 mol/l had no significant influence on the derived avg. activities.

From the investigation of the activity profiles it could be shown that $H_2$ accelerates the reaction start. Therein, higher activities as well as a stronger temperature increase at the beginning of the reaction were observed with increasing $H_2$ concentration. In contrast, the normalized activity profiles with the corresponding avg. activity showed no significant influence of $H_2$ on the deactivation behavior. Similar deactivation behavior was observed for the different $H_2$ concentrations.

Effect of different reaction conditions and catalyst injection conditions on polymer characteristics

Very similar tendencies of the influences of the different reaction conditions on the polymer characteristics were observed for both catalysts. In polymerizations carried out with prepolymerization, spherical polymer particles with a smooth particle surface were produced at all investigated temperatures. Without prepolymerization, not perfectly spherical particles with a rough particle surface were produced. Too high reaction temperatures (90°C) led to non-spherical particles and particle breakage into flakes occurred. Different $H_2$ concentrations showed no significant influence on the derived polymer morphologies. From sieving analysis an increase in the amount of fines could be observed with increasing reaction temperatures as well as with increasing $H_2$ concentrations. The bulk density therefore decreased with increasing reaction temperatures and $H_2$ concentrations, respectively. Furthermore, lower bulk densities were derived without prepolymerization compared to polymerizations carried out with prepolymerization. No clear trend of the reaction conditions on porosity could be observed. Therefore, an avg. porosity of the polymer particles of 25 % is assumed. Sorption measurements with monomer have shown a similar sorption behavior of polymer particles produced with and without prepolymerization at low and high reaction temperatures.

70

# 5 Kinetic modeling of the gas-phase polymerization with two different Ziegler-Natta catalysts

Kinetics of Ziegler-Natta polymerizations are very complex. Multiple active sites of the Ziegler-Natta catalyst, the morphology of the polymer particles, the phase equilibria as well as mass and heat transfer can influence the polymerization behavior.

Purpose of the modeling activities in this work is to develop a simplified phenomenological kinetic model approach for the quantitative correct description of the gas-phase polymerization of propylene for the investigated ZN catalysts and to determine the corresponding kinetic parameters. Therein, the model should be able to describe the influence of the different reaction conditions as well as the effect of the different catalyst injection conditions (prepolymerization) on the activity for both ZN catalysts. Also weight average molecular weights are predicted with the model.

For the derivation of the kinetic model approach, the experimentally derived activity profiles, the average activities as well as the determined weight average molecular weights shown in chapter 4.6 were analyzed. Modeling and parameter estimation were carried out using the simulation software gProms ModelBuilder (Process Systems Enterprise Ltd.).

In the following chapter 5.1, the derivation of the kinetic model as well as the involved mass and heat balances are described. The results of the parameter estimation and the comparison of simulated and experimental data are shown in chapter 5.2.

## 5.1 Derivation of the kinetic model

### 5.1.1 Model assumptions

With the aim to develop a simplified kinetic model approach, some simplifications are necessary. One main simplification of the model is that, even though ZN catalysts are known to have multiple active sites, only one kind of active sites is assumed. This assumption was made in order to have a simplified kinetic scheme and to reduce the number of kinetic constants which need to be estimated by fitting the experimental data.

The mass balances are derived for the polymer particles. As simplification, mass transfer limitations of the monomer into the polymer particle are neglected. Furthermore, no swelling of the polymer by the monomer is considered. The equilibrium monomer concentration within the polymer particle is calculated by using Henry's Law and the equation of Stern (chapter 5.1.8).

The effect of hydrogen on activity and molecular weight will be also implemented in the model. It is therefore assumed that the hydrogen concentration in the particle is equal to the hydrogen concentration in the gas-phase, no mass transfer limitations for hydrogen are considered. Average molecular weights are calculated using the method of moments (chapter 5.1.7).

Temperature dependencies of the rate constants are generally described by the Arrhenius equation. Therein, the temperature is based on the calculated particle temperature (chapter 5.1.6).

In the comparison of the experimental results of both ZN catalysts it could be shown that similar trends of the avg. activities were derived for both catalysts at the different reaction conditions. Also the activity profiles showed similar behaviors at the different reaction conditions. Following main conclusions can be summarized from the experimental findings:

- The activity of catalyst B is approx. double as the activity of catalyst A.
- The effect of high injection temperatures on activity is stronger for catalyst B.
- The weight average molecular weight and the hydrogen response on the activity are similar for both catalysts.
- A prepolymerization step improves the activities for both catalysts at higher reaction temperatures.
- The influence of injection temperature on activity is catalyst specific.

Based on the experimental results it is therefore assumed that the kinetic behaviors of both catalysts A and B can be described with the same model approach.

## 5.1.2 Derivation of the kinetic scheme

The derivation of the model approach is based on the experimentally observed activity profiles. The current activity of the catalyst over the reaction time is proportional to the overall reaction rate of the polymerization reaction. The overall reaction rate of the polymerization is proportional to the rate constant of the propagation reaction, the monomer concentration and the concentration of active sites according:

$$A \approx R_P \approx k_p \cdot c_M \cdot c_{p*} \cdot V_R \tag{5.1}$$

with $R_P$ as overall reaction rate of the polymerization, $k_p$ as propagation rate constant, $c_M$ as monomer concentration, $c_{p*}$ as concentration of the growing chains and $V_R$ as reaction volume.

In coordination polymerization, the concentration of the growing chains refers directly to the concentration of the active sites of the catalyst. During the polymerization, the reaction volume, which is the volume of the polymer phase, will increase. This volume increase would be therefore result in a decrease of the concentration of the active sites, the so-called dilution effect of the active sites. However, the moles n of active sites as product of concentration and volume only changes due to reaction, hence the moles are chosen as balancing quantity [91]:

$$A \approx R_P \approx k_p \cdot c_M \cdot n_{p*} \tag{5.2}$$

Formation of active sites

Active sites are formed via the reaction of the active component of the catalysts with the cocatalyst:

(I)    *Formation of active sites*:        $Ti + TEA \rightarrow Ti^*$        $k_{act}$

Herein, the rate constant $k_{act}$ is assumed to be temperature independent as no exact kinetic information of the reaction start could be derived from the experimental data.

For the calculations, the initial number of moles of the active sites has to be known. For heterogeneous catalysts, the total number of moles of active component can be determined from the weighted amount of catalyst, the weight fraction of the active component of the supported catalyst and its molecular weight. Furthermore, it is known that not all of the active component Ti is activated, e.g. due to incomplete activation by the cocatalyst, by poisoning with reactor impurities, deactivation reactions with functional groups on the support or steric hindrance by the support surface [7]. It is assumed that for a $TiCl_4/MgCl_2$ catalyst only 1 to 10 % of the Ti atoms are active for the polymerization [7]. For that purpose, a semi-empirical correction factor is implemented which describes the active fraction of Ti for the polymerization. The number of moles of active sites at the beginning of the reaction ($n_{Ti}^0$) is therefore calculated according following equation:

$$n_{Ti}^0 = \frac{w_{Ti} \cdot m_{Cat}}{MW_{Ti}} \cdot x_{activ} \tag{5.3}$$

with $m_{Cat}$ as weighted amount of catalyst, $w_{Ti}$ as weight fraction of the active component of the catalyst, $MW_{Ti}$ as molecular weight of the active component and with $x_{activ}$ as semi-empirical correction factor.

As described before, the same kinetic model approach can be used for both investigated ZN catalysts. In order to take the specifics of each catalyst into account, a different total amount of the active component at the beginning of the reaction ($n_{Ti}^0$), represented by catalyst specific values of $w_{Ti}$ and $x_{activ}$, is considered.

Chain initiation

In the chain initiation reaction, the activated Ti* reacts with the monomer and forms a polymer with chain length one.

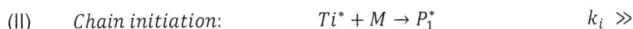

(II)    *Chain initiation*:        $Ti^* + M \rightarrow P_1^*$        $k_i \gg$

As the initiation reaction directly occurs at the beginning of the polymerization, it could not be determined experimentally by the kinetic measurements. For the model it is considered that the rate constant of the initiation reaction is very high. It is therefore not a rate limiting reaction step and will not be taken into account in the parameter estimation.

Chain propagation

The growth of the polymer chain occurs generally by the consumption of monomer in the propagation reaction step. The monomer is coordinated to the active site of the catalyst and inserted into the growing polymer with chain length n (see chapter 2.2.2 ). A polymer with the chain length (n+1) is formed.

(III)    *Propagation:*          $P_n^* + M \rightarrow P_{n+1}^*$          $k_p$

In the activity profiles it could be seen that there is a strong effect of the reaction temperature on the rate of the polymerization (4.6.4.2 and 4.6.5.2). An increase of the reaction temperature led to an increase of the reaction rate resulting in an increase of the (average) activity. The temperature dependency of the propagation reaction is therefore described by the Arrhenius equation:

$$k_p = k_p^0 \cdot e^{\left(-\frac{E_{A,p}}{RT}\right)}$$          (5.4)

With the pre-exponential factor $k_p^0$ and the activation energy $E_{A,p}$ of propagation, the reaction temperature $T$ and the universal gas constant $R$.

Influence of hydrogen on polymerization

In the experimental results it was shown that there is a strong effect of the hydrogen concentration on the activity level (chapter 4.6.4.4 and 4.6.5.4). With increasing $H_2$ concentration the activity increases until an activity plateau was reached for a certain $H_2$ concentration. Furthermore, it was observed that the presence of $H_2$ led to a faster reaction start. As discussed in chapter 2.3, one widely accepted explanation is that, due to a regioirregular insertion of the monomer into the growing chain, the so-called dormant sites are formed which are relatively unreactive for further propagation. A further monomer insertion is prevented or retarded. Additional reaction steps for the formation and reactivation of the dormant sites are therefore included based on the proposed elementary reactions steps by [62],[75],[82] (see chapter 2.2.2). In the model, the formation of the dormant chains is implemented as follows, wherein $P_n^*$ denotes the growing chains and $Y_n$ denotes the dormant chains.

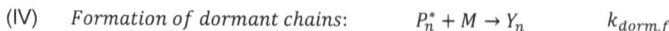

(IV)    *Formation of dormant chains:*          $P_n^* + M \rightarrow Y_n$          $k_{dorm,f}$

The inactive dormant sites can be reactivated by chain transfer with hydrogen releasing an active site for further propagation. As simplification, initiation reaction of the active site with monomer and further propagation is combined in one step. In the second step, the dormant chains are therefore reactivated by hydrogen forming growing chains again.

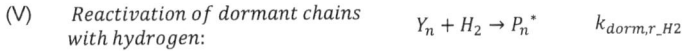

(V)  $Reactivation\ of\ dormant\ chains$ $with\ hydrogen$:  $\qquad$ $Y_n + H_2 \rightarrow P_n^*$ $\qquad$ $k_{dorm,r\_H2}$

In combination, these two reactions lead to a dynamic equilibrium of active and dormant chains as function of hydrogen concentration. However, this approach would lead to a strong decay in activity for polymerizations without hydrogen which was not observed experimentally. Therefore, an additional spontaneous reactivation reaction for the dormant chains is assumed in order to describe the polymerization reactions without hydrogen.

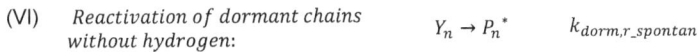

(VI)  $Reactivation\ of\ dormant\ chains$ $without\ hydrogen$:  $\qquad$ $Y_n \rightarrow P_n^*$ $\qquad$ $k_{dorm,r\_spontan}$

As no clear distinction between temperature dependency of propagation reaction and the formation and reactivation of dormant sites could be determined from the experimentally derived kinetic profiles, the rate constants for the dormant sites will be assumed as temperature independent.

Transfer reactions

Within a polymerization also chain transfer reactions occur. Therein, the growing polymer chain reacts to a death polymer chain and an active species. This active species will initiate another growing chain by monomer insertion (see chapter 2.2.2). In this model, it is assumed that all formed active species are similar and undergo instantaneously the initiation reaction with the monomer forming an active polymer with chain length one (according reaction II).

In general, hydrogen acts as chain transfer agent influencing the molecular weight of the polymer. For polymerizations with $H_2$, the calculation of the molecular weight is therefore mainly governed by the transfer reaction to hydrogen.

(VII)  $Transfer\ to\ hydrogen$:  $\qquad$ $P_n^* + H_2 \rightarrow D_n + P_1^*$ $\qquad$ $k_{tr,H2}$

For reactions without hydrogen, therefore without any transfer reaction, the molecular weight was calculated by magnitudes too high. Hence, further transfer reactions must be present. In the model, spontaneous chain transfer (by ß-hydride elimination) is considered as additional transfer reaction. Further transfer reactions, such as transfer to monomer and to cocatalyst, are neglected.

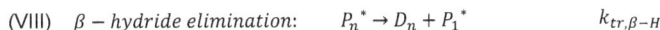

(VIII)  $\beta - hydride\ elimination$:  $\qquad$ $P_n^* \rightarrow D_n + P_1^*$ $\qquad$ $k_{tr,\beta-H}$

The transfer reactions are temperature dependent which is also described by the Arrhenius equation:

$$k_{tr,H2} = k_{tr,H2}^0 \cdot e^{\left(-\frac{E_{A,trH2}}{RT}\right)} \tag{5.5}$$

$$k_{tr,\beta-H} = k_{tr,\beta-H}^0 \cdot e^{\left(-\frac{E_{A,tr\beta-H}}{RT}\right)} \tag{5.6}$$

Deactivation

In the experimentally derived activity profiles, an activity loss during the polymerizations reaction was observed. It is generally accepted that this activity loss is due to the deactivation of the active sites. Furthermore, the deactivation can occur by reaction with other components ($H_2$, cocatalyst, monomer), by poisoning or spontaneously (see chapter 2.2.2).

In the experimental investigations, activity profiles derived at different reaction pressures (different monomer concentrations, respectively) were normalized with the corresponding monomer concentration (chapter 4.6.4.1). It could be shown that monomer concentration has no significant influence on the deactivation profile. Also for different $H_2$ concentrations (chapter 4.6.4.4), no influence on the deactivation profile was visible. The influence of cocatalyst on the deactivation profile was not measurable because always same amount of cocatalyst was used. Poisoning of the catalyst is neglected due to the excess of cocatalyst acting also as scavenger.

In the model, the decrease of activity over reaction time is therefore described by a spontaneous deactivation of the catalyst which is independent of the monomer concentration.

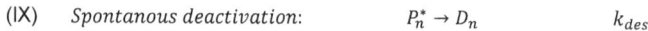

(IX)    *Spontanous deactivation:*        $P_n^* \rightarrow D_n$        $k_{des}$

As the activity profiles normalized with the corresponding avg. activity showed a similar deactivation behavior (chapter 4.6.4.2. and 4.6.5.2), the temperature dependency of the spontaneous deactivation is also described with the Arrhenius equation:

$$k_{des} = k_{des}^0 \cdot e^{\left(-\frac{E_{A,des}}{RT}\right)} \tag{5.7}$$

## 5.1.3  Influence of catalyst injection conditions – effect of prepolymerization

In chapter 4.6.4.3, the experimental results of the polymerization with and without prepolymerization were compared. It could be shown that prepolymerization has a significant influence on the activity especially at high reaction temperatures. Without prepolymerization, a strong decrease of activity at higher reaction temperatures was observed. For polymerizations carried out with prepolymerization much higher activities were reached. Comparing the different catalysts (chapter 4.6.5.3) it could be seen that the impact of prepolymerization depends on the catalyst activity at reaction temperature. The higher the catalyst activity the more critical is the effect of the catalyst injection temperature on the reached activity.

In order to describe the different behavior of the catalyst at the different injection conditions in the model, it has to be clarified how the differences in activity could be explained. Starting from the rate law (equation 5.1), the activity is proportional to the rate constant $k_p$, the equilibrium monomer concentration $c_M{}^*$ and the amount of active sites $c_p{}^*$. The injection conditions should affect neither the propagation rate constant nor the equilibrium monomer concentration in the polymer particle (see chapter 4.6.6.7). Therefore, the amount of active sites remains as major reason for the observed differences. The amount of active sites can change via formation or deactivation reactions. In chapter 4.6.4.3 and 4.6.5.3, respectively, it was shown that the deactivation of active sites over time is similar for all reactions, independent of the injection conditions.

A possible explanation could be that particle overheating especially at the beginning of the reaction is the major reason for the lower activities obtained for polymerizations without prepolymerization. In order to test this hypothesis, a simulation case study on particle overheating was carried out based on a first model approach. Therein, the maximum temperature increase in the polymer particle was calculated for polymerizations with and without prepolymerization for the investigated reaction temperatures in the initial stage of the main polymerization reaction. The estimated particle temperature differences (between particle temperature and surrounding gas-phase temperature) are shown in Figure 51. As it can be seen, there is a significant temperature increase when no prepolymerization step is applied. The particle temperature strongly increases with increasing reaction temperature (gas-phase). When using a prepolymerization step, the particle temperature only slightly increases with increasing reaction temperature.

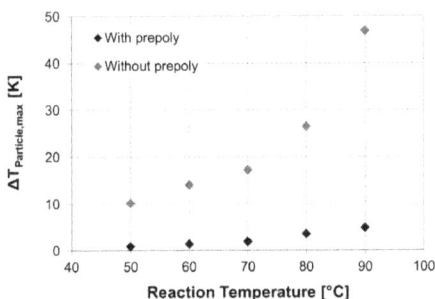

Figure 51: Simulation case study on particle overheating – polymerization with and without prepolymerization

The much higher particle temperature could be therefore lead to a stronger deactivation of the active sites. Based on the case study, it is therefore assumed that particle overheating might cause an initial fast thermal deactivation of the active sites leading to lower activities in case of polymerizations without prepolymerization.

77

In the model, this is taken into account via:

- calculating the particle growth and particle temperature during polymerization and
- considering a further strong temperature dependent deactivation reaction for active component:

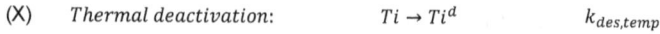

$$(X) \quad Thermal\ deactivation: \quad Ti \rightarrow Ti^d \quad\quad k_{des,temp}$$

The strong temperature dependency of the thermal deactivation reaction is also described by the Arrhenius equation:

$$k_{des,temp} = k^0_{des,temp} \cdot e^{\left(-\frac{E_{A,des,temp}}{RT}\right)} \tag{5.8}$$

Finally, as the particle temperature is the actual temperature at the active sites, the temperature in the Arrhenius equations for the considered elementary reaction steps (equations 5.4 to 5.8) is based on the particle temperature. The calculation of the particle temperature is described in chapter 5.1.6.2.

### 5.1.4 Final kinetic scheme

The final simplified kinetic scheme for the gas-phase polymerization is shown in Figure 52. The same kinetic scheme is used to describe the polymerizations with both investigated ZN catalysts. Therein, the effects of reaction temperature and hydrogen concentration on activity and molecular weight as well as the effect of the different catalyst injection conditions are considered. The effect of reaction pressure, respectively monomer concentration, is taken into account by calculating the equilibrium monomer concentration in the polymer particle and is described in chapter 5.1.8.

| | | | |
|---|---|---|---|
| Formation active sites | $Ti + TEA \rightarrow Ti^*$ | $r_{act} = k_{act} \cdot c_{Ti}$ | (I) |
| and initiation | $Ti^* + M \rightarrow P_1^*$ | $k_i$ considered to be very fast | (II) |
| Chain propagation | $P_n^* + M \rightarrow P_{n+1}^*$ | $r_p = k_p \cdot c_{Pn*} \cdot c_M$ | (III) |
| Formation dormant sites | $P_n^* + M \rightarrow Y_n$ | $r_{dorm,f} = k_{dorm,f} \cdot c_{Pn*} \cdot c_M$ | (IV) |
| Reactivation dormant sites | $Y_n + H_2 \rightarrow P_n^*$ | $r_{dorm,r\_H2} = k_{dorm,r\_H2} \cdot c_{Yn} \cdot c_{H2}$ | (V) |
| | $Y_n \rightarrow P_n^*$ | $r_{dorm,r\_spontan} = k_{dorm,r\_spontan} \cdot c_{Yn}$ | (VI) |
| Chain transfer[1] | $P_n^* + H_2 \rightarrow D_n + P_1^*$ | $r_{tr,H2} = k_{tr,H2} \cdot c_{Pn*} \cdot c_{H2}^{0.5}$ | (VII) |
| | $P_n^* \rightarrow D_n + P_1^*$ | $r_{tr,\text{ß}-H} = k_{tr,\text{ß}-H} \cdot c_{Pn*}$ | (VIII) |
| Spontaneous deactivation | $P_n^* \rightarrow D_n$ | $r_{des} = k_{des} \cdot c_{Pn*}$ | (IX) |
| Thermal deactivation | $Ti \rightarrow Ti^d$ | $r_{des,Temp} = k_{des,Temp} \cdot c_{Ti}$ | (X) |

**Figure 52: Kinetic scheme for the gas-phase polymerization with ZN catalyst**

## 5.1.5  Mass balances of the reaction components

Based on the developed kinetic scheme, the following mass balance for the growing polymer chains is derived:

$$\frac{dn_{Pn*}}{dt} = V_R \cdot R_{Pn*} = V_R \cdot \left(r_{act} + r_{dorm,r\_H2} + r_{dorm,r\_spontan} - r_{dorm,f} - r_{des}\right) \tag{5.9}$$

Herein, growing chains or active sites, respectively, are formed via the activation (and immediately initiation) reaction, the reactivation of the dormant sites with hydrogen and spontaneous and are consumed by the formation of dormant sites and the spontaneous deactivation of the catalyst. Propagation and transfer reactions are not included as they do not change the amount of growing chains itself. As described in chapter 5.1.2, the concentration of the active sites and dormant sites, respectively, will be change during the reaction due to the growth of the polymer particle (dilution effect). Therefore, the number of moles as a product of reaction volume and concentrations of active sites and dormant sites is used for the calculations. Together with the rate expressions given in Figure 52, the resulting mass balance for the growing polymer chains is:

---

[1] It was found in literature [6],[27],[101] and also confirmed with the model that the reaction order of hydrogen for the transfer reaction with MgCl$_2$-supported catalysts is 0.5.

$$\frac{dn_{Pn*}}{dt} = k_{act} \cdot n_{Ti} + k_{dorm,r_{H2}} \cdot n_{Yn} \cdot c_{H2} + k_{dorm,r_{spontan}} \cdot n_{Yn} - k_{dorm,f} \cdot n_{Pn*} \cdot c_{C3} \tag{5.10}$$
$$- k_{des} \cdot n_{Pn*}$$

The mass balances for the other components, the active component $n_{Ti}$, the dormant chains $n_{Yn}$ as well as dead polymer chains $n_{Dn}$, are obtained in the same manner:

$$\frac{dn_{Ti}}{dt} = -k_{act} \cdot n_{Ti} - k_{des,Temp} \cdot n_{Ti} \tag{5.11}$$

$$\frac{dn_{Yn}}{dt} = k_{dorm,f} \cdot n_{Pn*} \cdot c_{C3} - k_{dorm,r_{H2}} \cdot n_{Yn} \cdot c_{H2} - k_{dorm,r\_spontan} \cdot n_{Yn} \tag{5.12}$$

$$\frac{dn_{Dn}}{dt} = k_{tr,H2} \cdot c_{Pn*} \cdot c_{H2}^{0.5} + k_{tr,\beta-H} \cdot c_{Pn*} + k_{des} \cdot c_{Pn*} \tag{5.13}$$

The initial number of moles of active component $n_{Ti}^0$ is calculated according equation (5.3). The initial number of moles of growing chains, dormant chains and dead chains is equal to zero.

$$n_{Pn*}^0 = n_{Yn}^0 = n_{Dn}^0 = 0 \tag{5.14}$$

The amount of polymer produced over reaction time is equal to the total consumed monomer, expressed as overall monomer consumption rate $R_M$ times reaction volume $V_{Pol}$ (polymer phase) and molecular weight of the monomer $MW_{C3}$:

$$\frac{dm_{Pol}}{dt} = -R_M \cdot V_{Pol} \cdot MW_{C3} \qquad \left[\frac{g_{Pol}}{s}\right] \tag{5.15}$$

with the initial condition:

$$m_{Pol}^0 = 0 \tag{5.16}$$

According to the reaction scheme, monomer is consumed during propagation and by the formation of the dormant sites. As the produced amount of polymer from dormant chains is not significant (most of the dormant chains will be reactivated by $H_2$ or spontaneously), it is neglected for further calculations.

$$-R_M = R_P = r_p + r_{dorm,f} \approx r_p \tag{5.17}$$

The current activity can be therefore calculated by following equation:

$$A = \frac{dm_{Pol}}{dt} \cdot \frac{1}{m_{Cat}} = \frac{r_p \cdot V_{pol} \cdot MW_{C3}}{m_{Cat}} \qquad \left[\frac{g_{PP}}{g_{Cat} \cdot s}\right] \tag{5.18}$$

With the expression of the rate law for $r_p$, the consideration of the dilution effect and the conversion of the units, the current activity is finally calculated according following equation:

$$A = \frac{k_p \cdot n_{Pn*} \cdot c_{C3} \cdot MW_{C3}}{m_{Cat}} \cdot \frac{3600\frac{s}{h}}{1000\frac{g}{kg}} \qquad \left[\frac{kg_{PP}}{g_{Cat} \cdot h}\right] \tag{5.19}$$

## 5.1.6 Particle growth and particle heat balance

As described in the derivation of the kinetic scheme, a further strong temperature dependent thermal deactivation reaction was implemented in order to describe the influence of the catalyst injection conditions (prepolymerization as well as effect of high reaction temperatures) on the activity. It is assumed that particle overheating at the beginning of the reaction is the major reason for the lower activities obtained when no prepolymerization step is applied. In order to describe the temperatures within the polymer particle, particle growth and particle heat balances have to be implemented into the model.

For the calculations, a quasi-homogeneous particle model is assumed (Figure 53). Herein, the catalyst and the polymer particle build one phase. The different particle morphology is not considered. The balancing domain is the polymer particle. Therein, no mass transfer limitations as well as no swelling of the polymer particle by the monomer are assumed. The equilibrium monomer concentration inside the polymer particle is calculated by Henry's law and Stern equation (chapter 5.1.8). Furthermore, heat transfer limitations from the particle surface to the surrounding gas-phase are considered. Herein, the accumulated heat in the polymer particle is equal to the produced heat by the exothermal polymerization reaction reduced by the removed heat by heat transfer via the particle surface. The heat of sorption of the sorbed monomer is not considered separately but is instead included in the heat of reaction.

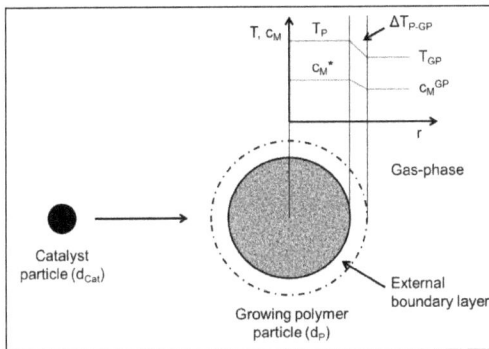

Figure 53: Scheme of growing polymer particle – mass and heat balance [99]

In the following, the derivations of the mass and heat balances for one polymer particle as balancing domain are described.

### 5.1.6.1 Particle growth

Similar to the mass balance for the total polymer mass (equation 5.16), the growth of one polymer particle can be described by following mass balance:

$$\frac{dm_P}{dt} = \frac{d}{dt}\left(\frac{\pi d_P^3}{6} \cdot (1 - \varepsilon_P) \cdot \rho_P\right) = -\frac{R_M \cdot V_{Pol} \cdot MW_{C3}}{N} \tag{5.20}$$

with $m_P$ as mass, $d_P$ as diameter, $\varepsilon_P$ as porosity and $\rho_P$ as density of the polymer particle, $V_{Pol}$ as polymer volume and $N$ as total number of particles.

The number of particles can be calculated from the weighted mass of catalyst related to the calculated mass of one catalyst particle:

$$N = \frac{m_{Cat}}{\left(\frac{\pi d_{Cat}^3}{6} \cdot (1 - \varepsilon_{Cat}) \cdot \rho_{Cat}\right)} \tag{5.21}$$

with $m_{Cat}$ as mass, $d_{Cat}$ as diameter, $\varepsilon_{Cat}$ as porosity and $\rho_{Cat}$ as density of the catalyst.

Substituting N in equation (5.20) with equation (5.21) and simplifying lead to the following equation describing the particle growth in terms of the particle diameter over time:

$$\frac{dd_P}{dt} = -\frac{R_M \cdot V_P \cdot MW_{C3}}{3 \cdot m_{Cat}} \cdot \frac{d_{Cat}^3 \cdot (1 - \varepsilon_{Cat}) \cdot \rho_{Cat}}{d_P^2 \cdot (1 - \varepsilon_P) \cdot \rho_P} \tag{5.22}$$

In a next step, a growing factor g is defined which describes the ratio of polymer diameter to catalyst diameter:

$$g = \frac{d_P}{d_{Cat}} \tag{5.23}$$

As described in the previous chapter, $-R_M$ can be assumed to be equal to the reaction rate of propagation (equation 5.17). With combination of equations (5.22) and (5.23), the growth of the polymer particle can be calculated at any reaction time according following equation:

$$\frac{dg}{dt} = \frac{r_p \cdot V_P \cdot MW_{C3}}{3 \cdot m_{Cat} \cdot g^2} \cdot \frac{(1 - \varepsilon_{cat})}{(1 - \varepsilon_P)} \cdot \frac{\rho_{Cat}}{\rho_P} \tag{5.24}$$

Rearranging with the definition of the activity (equation (5.18), the growth of the polymer particle is finally calculated by:

$$\frac{dg}{dt} = \frac{A}{3 \cdot g^2} \cdot \frac{(1 - \varepsilon_{cat})}{(1 - \varepsilon_P)} \cdot \frac{\rho_{Cat}}{\rho_P} \tag{5.25}$$

For the catalyst density, the density of the support is assumed (see Table 5 chapter 4.2). The catalyst porosity is given with 40 %. An avg. polymer density of 0.905 g/ml is used for the calculations, which was determined by means of helium pycnometry for the different polymer powders (see Appendix 9.1.2.3). An avg. porosity of the polymer of 25 % was determined (see chapter 4.6.6.6) and is used for the calculations.

In order to describe the different catalyst injection conditions, the growing factor at the beginning of the reaction needs to be defined. In case of polymerizations carried out without prepolymerization step, the growing factor is equal to one as size of the polymer particle and catalyst particle are the same. For polymerizations carried out with prepolymerization step, it is assumed that the reaction starts directly at the main polymerization conditions with an already prepolymerized catalyst. The growing factor is therefore greater-than one. The growing factor was determined from an additional prepolymerization experiment with catalyst B (see Appendix 9.1.3). Herein, only the prepolymerization step was carried out from 40°C to 80°C and stopped directly when 80°C were reached. For the prepolymer, a Malvern-analysis was carried out and a $D_{50}$ of 959 μm was determined. This corresponds to a growing factor g of 17.4.

The chosen initial conditions for g in the model are therefore:

- for polymerizations without a prepolymerization: $g^0 = 1$
- for polymerizations with prepolymerization: $g^0 = 18$

As the activity of catalyst A is lower compared to the activity of catalyst B, a slightly lower g would be expected. As simplification, the same growing factor is assumed for all polymerizations with prepolymerization as well as for both catalysts.

## 5.1.6.2 Particle heat balance

In order to calculate the current particle temperature, a heat balance is derived for the polymer particle. Therein, heat produced within the polymer particle and heat removed via the particle surface have to be considered.

In the model, it is assumed that heat is mainly released by the exothermal polymerization reaction. Heat generated by the sorption of the monomer into the polymer particle is included in the heat of reaction. As the heat conductivity in the polymer particle (0.17-0.22 W/(m·K) [136]) is higher than the heat conductivity in the gas-phase (0.022-0.03 W/(m·K)[132]), the main resistance for heat transport will be between particle surface and gas-phase (in the particle boundary layer). Therefore, no temperature gradient within the polymer particle is considered and a mean particle temperature is used for calculations. This assumption was already confirmed by Floyd et al.[96] and Hutchinson and Ray [98] or e.g. in simulations of Bartke [91] and Kröner [80].

From that consideration, following global heat balance is assumed for the polymer particle:

$$\dot{Q}_{accu} = \dot{Q}_{chem} - \dot{Q}_{removed} \qquad (5.26)$$

Therein, the heat accumulated in the polymer particle is:

$$\dot{Q}_{accu} = \frac{d\left(m_P \cdot c_{p,P} \cdot \bar{T}_P\right)}{dt} \qquad (5.27)$$

with $c_{p,P}$ as specific heat capacity of the particle (polypropylene) and $\bar{T}_P$ as mean particle temperature.

The heat produced by chemical reaction per polymer particle is described by:

$$\dot{Q}_{Chem} = \frac{-R_M \cdot V_{Pol} \cdot (-\Delta_R H)}{N}$$

(5.28)

with $\Delta_R H$ as reaction enthalpy and $N$ as number of particles according equation 5.21. The reaction enthalpy of the propylene polymerization from gaseous to solid state is 104 kJ/mol [137].

The removed heat by heat transfer from particle to gas-phase is described by the heat transition between particle surface and gas-phase:

$$\dot{Q}_{removed} = k \cdot A_P \cdot \Delta T = \alpha \cdot A_P \cdot (\bar{T}_P - T_G)$$

(5.29)

with $k$ as heat transfer coefficient, $\alpha$ heat transition coefficient, $A_P$ as particle surface and $T_G$ as temperature of the gas-phase or surrounding monomer, respectively.

The heat transition coefficient $\alpha$ can be calculated by means of the $Nu$ number, the heat conductivity coefficient of the gas-phase (monomer) $\lambda_G$ and the particle diameter $d_P$ as characteristic length:

$$\alpha = \frac{Nu \cdot \lambda_G}{d_P}$$

(5.30)

The particle surface $A_P$ is calculated by:

$$A_P = \pi d_P^2$$

(5.31)

Rearranging the global heat balance and with consideration of equation (5.20) for $m_P$, equation (5.21) for $N$ and equations (5.22) - (5.24) for $d_P$, following equation is derived for the mean particle temperature $\bar{T}_P$:

$$\frac{d\bar{T}_P}{dt} = -\frac{R_M \cdot V_{Pol} \cdot MW_{C3}}{m_{cat} \cdot g^3} \cdot \frac{(1 - \varepsilon_{cat})}{(1 - \varepsilon_P)} \cdot \frac{\rho_{Cat}}{\rho_P} \cdot \left( \frac{(-\Delta_R H)}{MW_{C3} \cdot c_{p,P}} - \bar{T}_P \right)$$
$$- \frac{6 \cdot Nu \cdot \lambda_G}{c_{p,P} \cdot g^2 \cdot d_{Cat}^2 \cdot (1 - \varepsilon_P) \cdot \rho_P} \cdot \Delta T$$

(5.32)

Also here, the superposition of the monomer consumption rate can be assumed to be equal to the reaction rate of propagation (see description to equation 5.17).

Together with consideration of the definition of the activity according to the equation (5.18), $\bar{T}_P$ can be finally calculated by:

$$\frac{d\bar{T}_P}{dt} = \frac{A}{g^3} \cdot \frac{(1 - \varepsilon_{cat})}{(1 - \varepsilon_P)} \cdot \frac{\rho_{Cat}}{\rho_P} \cdot \left( \frac{(-\Delta_R H)}{MW_{C3} \cdot c_{p,P}} - \bar{T}_P \right)$$
$$- \frac{6 \cdot Nu \cdot \lambda_G}{c_{p,P} \cdot g^2 \cdot d_{Cat}^2 \cdot (1 - \varepsilon_P) \cdot \rho_P} \cdot (\bar{T}_P - T_G)$$

(5.33)

At t=0, the particle temperature is equal to the gas-phase temperature:

$$\bar{T}_P^0 = T_G \tag{5.34}$$

For the calculation of the $Nu$ number several semi-empirical correlations exist. Floyd et al. [97] have shown that for gas-phase polymerizations the Ranz-Marshall correlation can be used for calculations (see also Hutchinson and Ray [98],[138]). It describes the particle-fluid heat transfer for a single sphere in a fluid medium (here monomer) which moves with a relative velocity $u$.

The Ranz-Marshall correlation is given by [139]:

$$Nu = 2 + 0.6 \cdot Re^{1/2} \cdot Pr^{1/3} \tag{5.35}$$

Therein, the Reynolds number $Re$ and Prandtl number $Pr$ are defined as:

$$Re = \frac{\rho_{C3} \cdot u \cdot d_P}{\eta_{C3}} \tag{5.36}$$

$$Pr = \frac{\eta_{C3} \cdot c_{p,C3}}{\lambda_G} \tag{5.37}$$

with $\rho_{C3}$ as density of propylene, $\eta_{C3}$ as dynamic viscosity of propylene, $u$ as relative velocity and $c_{p,C3}$ as specific heat capacity of propylene. The relative velocity $u$ is assumed to have a value of 0.02 m/s, which is according to Floyd et al. [97] appropriate for stirred bed reactors.

## 5.1.7 Determination of average molecular weights using the method of moments

In general, during polymerization a non-uniform product is formed consisting of a mixture of molecules of different chain lengths and therefore different molecular weights. For the calculation of a complete molecular weight distribution (MWD), mass balances for each polymer chain with chain length n for each polymer species have to be solved. Several special methods for the calculation of the MWD are available which differ in their information content and its numerical effort. A good overview of the several methods is given by Soares [76] and Bartke [140]. One of the most common and simple approach to calculate the MWD is the method of moments [15],[75],[76],[77]. Herein, only averages of the distribution - the number average and weight average molecular weight ($M_n$ and $M_w$) as well as the polydispersity index (PDI) - are derived.

In general, the $k^{th}$ moment of a species S is described by:

$$S_k = \sum_{n=1}^{\infty} n^k \cdot c_{Sn} = \sum_{n=1}^{\infty} n^k \cdot n_{Sn} \tag{5.38}$$

wherein the exponent $k$ is 0, 1 and 2 according to the zeroth, first and second moment, $c_S$ and $n_S$ are the concentration and number of moles, respectively, of the species $S$ and $n$ is the chain length.

The three moment equations have to be calculated for all polymer species present in the system: the growing chains (Q), the dormant chains (R) and the death chains (D), see Table 16.

| Moment | Growing chains | Dormant chains | Death chains |
|--------|----------------|----------------|--------------|
| $0^{th}$ | $Q_0 = \displaystyle\sum_{n=1}^{\infty} n_{Pn*}$ | $R_0 = \displaystyle\sum_{n=1}^{\infty} n_{Yn}$ | $D_0 = \displaystyle\sum_{n=2}^{\infty} n_{Dn}$ |
| $1^{st}$ | $Q_1 = \displaystyle\sum_{n=1}^{\infty} n \cdot n_{Pn*}$ | $R_1 = \displaystyle\sum_{n=1}^{\infty} n \cdot n_{Yn}$ | $D_1 = \displaystyle\sum_{n=2}^{\infty} n \cdot n_{Dn}$ |
| $2^{nd}$ | $Q_2 = \displaystyle\sum_{n=1}^{\infty} n^2 \cdot n_{Pn*}$ | $R_2 = \displaystyle\sum_{n=1}^{\infty} n^2 \cdot n_{Yn}$ | $D_2 = \displaystyle\sum_{n=2}^{\infty} n^2 \cdot n_{Dn}$ |

In order to describe the moments in dependency of the time, differential equations for each moment of each polymer species have to be derived:

$$\frac{dS_k}{dt} = \sum_{n=1}^{\infty} n^k \cdot \frac{dn_{Sn}}{dt} \tag{5.39}$$

The following set of differential equation is finally achieved for the growing, dormant and dead polymer chains from the mass balances according the kinetic scheme given in Figure 52:

*Growing chains:*

$$\frac{Q_0}{dt} = k_{act} \cdot n_{Ti} + \left(k_{dorm,r1} \cdot c_{H2} + k_{dorm,r2}\right) \cdot R_0 - \left(k_{dorm,f} \cdot c_{C3} + k_{des}\right) \cdot Q_0 \tag{5.40}$$

$$\frac{Q_1}{dt} = k_{act} \cdot n_{Ti} + k_p \cdot c_{C3} \cdot Q_0 + \left(k_{dorm,r1} \cdot c_{H2} + k_{dorm,r2}\right) \cdot R_1$$
$$+\left(k_{tr,H2} \cdot c_{H2}^{0.5} + k_{tr,\beta-H}\right) \cdot (Q_0 - Q_1) - \left(k_{dorm,f} \cdot c_{C3} + k_{des}\right) \cdot Q_1 \tag{5.41}$$

$$\frac{Q_2}{dt} = k_{act} \cdot n_{Ti} + k_p \cdot c_{C3} \cdot (2 \cdot Q_1 + Q_0) + \left(k_{dorm,r1} \cdot c_{H2} + k_{dorm,r2}\right) \cdot R_2$$
$$+\left(k_{tr,H2} \cdot c_{H2}^{0.5} + k_{tr,\beta-H}\right) \cdot (Q_0 - Q_2) - \left(k_{dorm,f} \cdot c_{C3} + k_{des}\right) \cdot Q_2 \tag{5.42}$$

*Dormant chains:*

$$\frac{R_0}{dt} = k_{dorm,f} \cdot c_{C3} \cdot Q_0 - \left(k_{dorm,r1} \cdot c_{H2} + k_{dorm,r2}\right) \cdot R_0 \tag{5.43}$$

$$\frac{R_1}{dt} = k_{dorm,f} \cdot c_{C3} \cdot Q_1 - \left(k_{dorm,r1} \cdot c_{H2} + k_{dorm,r2}\right) \cdot R_1 \tag{5.44}$$

$$\frac{R_2}{dt} = k_{dorm,f} \cdot c_{C3} \cdot Q_2 - \left(k_{dorm,r1} \cdot c_{H2} + k_{dorm,r2}\right) \cdot R_2 \tag{5.45}$$

*Dead chains:*

$$\frac{D_0}{dt} = \left(k_{tr,H2} \cdot c_{H2}^{0.5} + k_{tr,\beta-H} + k_{des}\right) \cdot Q_0 \tag{5.46}$$

$$\frac{D_1}{dt} = \left(k_{tr,H2} \cdot c_{H2}^{0.5} + k_{tr,\beta-H} + k_{des}\right) \cdot Q_1 \tag{5.47}$$

$$\frac{D_2}{dt} = (k_{tr,H2} \cdot c_{H2}^{0.5} + k_{tr,\beta-H} + k_{des}) \cdot Q_2$$

(5.48)

The weight average and number average molecular weight as well as PDI can be calculated from the moments at any time according to the following equations:

$$\bar{M}_n = MW_{C3} \cdot \frac{D_1}{D_0}$$

(5.49)

$$\bar{M}_w = MW_{C3} \cdot \frac{D_2}{D_1}$$

(5.50)

$$PDI = \frac{\bar{M}_w}{\bar{M}_n} = \frac{D_2 \cdot D_0}{D_1^2}$$

(5.51)

## 5.1.8 Calculation of equilibrium monomer concentration in the polymer particle

For the calculation of the several balances in the model, the equilibrium monomer concentration in the polymer particle needs to be calculated. As a quasi-homogeneous model is assumed, wherein mass transfer limitations as well as swelling are neglected (see chapter 5.1.6) and monomer concentration in gas-phase is not that high as e.g. in liquid phase, the equilibrium monomer concentration within the polymer particle is calculated by Henry's law:

$$c_M^* = k_H \cdot p_M$$

(5.52)

with $c_M^*$ as equilibrium monomer concentration, $p_M$ as monomer pressure and $k_H$ as Henry constant.

Stern et al. [141] proposed a correlation for the calculation of the Henry constant for several gases and vapors (including low molecular weight hydrocarbons) in a semi-crystalline polymer describing the dependency of the Henry constant from the temperature and the critical temperature of the penetrant. Hutchinson and Ray [99] used the correlation of Stern to fit literature data of several hydrocarbons including propylene in semi-crystalline polymer and proposed the following (more defined) correlation:

$$\log k_H = -2.38 + 1.08 \cdot \left(\frac{T_c}{T}\right)^2$$

(5.53)

with $T_c$ as critical temperature of the olefin and T as reaction temperature.

It has to be mentioned that Stern et al. [141] also proposed a correlation at which pressure the solubility calculated by Henry's law deviates by 5 % due to plasticizing effects by the penetrant:

$$log\left(\frac{p_{dev}}{p_c}\right) = 3.025 - 3.50\left(\frac{T_c}{T}\right)$$

(5.54)

wherein $p_c$ is the critical pressure and $p_{dev}$ is the pressure at which a deviation of 5 % occurs.

For example, for a polymerization carried out at 70°C and 27.5 bar, $p_{dev}$ would become 9.3 bar. Nevertheless, Kröner [80] showed in sorption measurements of propylene in semi-crystalline polypropylene carried out at higher pressures up to 25 bar that deviations are in an acceptable range of approx. less than 10 %. Therefore, and based on the simplifications made within this model, Henry's law is appropriate to calculate the equilibrium monomer concentration within the polymer particle at the investigated pressure range.

### 5.1.9 Model implementation and parameter estimation

Modeling and parameter estimation were carried out using the software gProms ModelBuilder V4.2.0 (Process Systems Enterprise Ltd., documentation see [142]).

In the "model" section, the set of differential equations of the components derived from the kinetic scheme, the moment equations of each species as well as the equations for the calculation of the particle growth, particle temperature, monomer concentration, avg. molecular weight and the activity were implemented. Data of the physical properties such as density, specific heat capacity, heat conductivity coefficient and dynamic viscosity of propylene were taken from the NIST database [132]. Catalyst relevant data such as the amount of active component on the catalyst particle were provided by the cooperation partner and are confidential. Data of the polymer density, porosity and crystallinity were taken from the analytical measurements.

In the "process" section, the kinetic constants (rate constants or pre-exponential factor and activation energy, respectively), the reaction conditions (T, p, $m_{Cat}$, $c_{H2}$, g), the catalyst parameters ($w_{Ti}$, $x_{active}$) and the initial conditions of the differential equations were defined.

The parameter estimation was carried out with the integrated gEst parameter estimation tool. Herein, initial values of the rate constants needed to be estimated were entered together with a lower and upper bound for the estimation. The parameter estimation in gPROMS is based on the Maximum Likelihood formulation [142]. In order to take into account the uncertainties of the measurement, a statistical standard deviation can be defined in the gPROMS parameter estimation tool. For the parameter estimation, a constant relative variance of 0.1 was chosen which corresponds to an experimental error of the measured data of 10%.

For the parameter estimation, the experimental data (activity and weight avg. molecular weight) were implemented in the "experiments performed" section. In order to avoid too many data points, the measured activity-time profiles were reduced by using the current activity at selected times.

As shown in the experimental results, both catalysts show a similar behavior in terms of activity and $M_W$ at different reaction conditions including also the effect of different injection conditions (prepolymerization, high reaction temperatures). First parameter estimations and simulations have shown that kinetic parameters are also similar for both catalysts; the major difference between both catalysts is the amount of the active component. Based on this finding, the parameter estimation was carried out together with all experimental results of both catalysts to determine final kinetic constants valid for both catalysts in the investigated parameter range.

In general, the estimation of the kinetic parameters was performed by fitting the model to the experimentally derived activity-time profiles and molecular weights, respectively. Due to the relative high number of the kinetic constants (see kinetic scheme chapter 5.1.4), the estimation was performed stepwise. Firstly, rate constants of initiation, propagation and deactivation reaction were estimated by fitting the model with the experimental activity-time profiles for the different reaction temperatures at one hydrogen concentration.

The influence of hydrogen on the activity profiles could be clearly shown in the experiments and is considered in the model by the dormant site theory. Therefore, in a second step, the rate constants for the formation and reactivation of dormant sites were estimated. Herein, activity profiles of experiments with and without $H_2$ of all reaction temperatures were used. In both steps, rate constants of the transfer reaction were not taking into account as they have no influence on the number of active sites, and therefore on the activity-time profiles used for the estimation.

In a third step, the rate constants of the transfer reactions (transfer to $H_2$ and ß-hydride elimination) were estimated by fitting the model to the experimental weight average molecular weights.

The effect of prepolymerization is considered in the model by estimating the current particle temperature and particle growth. For polymerizations with prepolymerization, a starting growth factor of the prepolymer of $g^0$=18 was chosen, for polymerizations without prepolymerization $g^0$ was set as per definition to 1 (see chapter 5.1.6.1). In order to describe the effect of the high temperatures and the impact of the prepolymerization on activity, the additional thermal deactivation was implemented. Temperatures used in the Arrhenius equations for the several elementary reaction steps are based on the particle temperature.

In a fourth step, the parameter estimation for the thermal deactivation was therefore carried out with experiments at higher reaction temperatures from 70°C to 90°C, with and without pre-polymerization. The spontaneous deactivation was corrected by repeating the parameter estimation only with reactions at 50°C and 60°C.

As the estimation of the parameters is an iterative process, each step was checked carefully on plausibility before proceeding to the next step. At the end, the final set of parameters was cross checked and the parameter estimation was repeated again until the best fit with the derived kinetic parameters was derived.

## 5.2 Results kinetic modeling of the gas-phase polymerization with Ziegler-Natta catalysts

### 5.2.1 Results parameter estimation

As described earlier, based on the experimental findings, the kinetic behavior of both catalysts can be described with the same model approach. Also first estimations showed that the kinetic parameters are similar for both catalysts. Thus, the same kinetic parameters for the several elementary reaction steps were estimated; the differences between the two catalysts are only described with the different amount of active sites available at the beginning of the reaction. As catalyst B has a higher activity compared to catalyst A, a higher value of the fraction of the initial polymerization active component (product of $w_{Ti}$ and $x_{active}$) is derived. The following set of kinetic parameters was finally obtained from the parameter estimation for both catalysts A and B:

Table 17: Results parameter estimation for the gas-phase polymerization with different ZN catalysts

| Parameter | Cat A | Cat B | Unit |
|---|---|---|---|
| $w_{Ti} \cdot x_{acti}$ | 2.17 | 3.70 | wt.% |
| $k_{act}$ | 0.05 | | 1/s |
| $k_p$ (70 °C) | 16520 | | l/(mol·s) |
| - $E_{A,p}$ | 56.2 | | kJ/mol |
| $k_{tr \, ß-H}$ (70 °C) | 3.6 | | 1/s |
| - $E_{A,tr \, ß-H}$ | 90.4 | | kJ/mol |
| $k_{tr \, H2}$ (70 °C) | 41.6 | | l/(mol·s) |
| - $E_{A,tr \, H2}$ | 45.4 | | kJ/mol |
| $k_{dorm,f}$ | 0.016 | | l/(mol·s) |
| $k_{dorm,r1}$ | 11.85 | | l/(mol·s) |
| $k_{dorm,r2}$ | 0.018 | | 1/s |
| $k_{des}$ (70 °C) | $2.14 \cdot 10^{-4}$ | | 1/s |
| - $E_{A,des}$ | 17.6 | | kJ/mol |
| $k_{des \, Temp}$ (70 °C) | $3.24 \cdot 10^{-3}$ | | l/(mol·s) |
| - $E_{A,des \, Temp}$ | 181.3 | | kJ/mol |

A comparison of the estimated kinetic parameter with literature data is difficult as catalysts (catalyst generations) and reaction conditions (gas or liquid phase, temperature, hydrogen concentration) are different. General reference data for the polymerization of propylene using ZN catalysts can be found in [28],[83],[57],[62],[63],[92],[143]. As example, for kinetic studies of the GP polymerization, Soares and Hamielec [28] estimated for a non-supported, high active ZN catalyst an $E_{A,p}$ of 57.7 kJ/mol and 65.3 kJ/mol for reactions without and with hydrogen, respectively. Patzlaff [92] estimated for high active ZN catalysts of different supports $E_{A,p}$ in the range from 28 to 84 kJ/mol. Samson et al. [63] estimated for an high active $MgCl_2$ supported catalyst for propagation an $E_{A,p}$ of 76.2 kJ/mol and for deactivation an $E_{A,d}$ of 50.6 kJ/mol (T = 22 - 52°C).

The here estimated activation energy for the thermal deactivation is rather high as it covers the strong decrease of the catalyst activity at high reaction temperatures as well as the effect of prepolymerization (different catalyst injection conditions). The activation energy of the spontaneous deactivation $E_{A,des}$ is therefore lower compared to literature data.

## 5.2.2 Comparison simulated and experimentally derived data – ZN catalyst B

### 5.2.2.1 Influence of reaction temperature

Figure 54 shows the simulated activity profiles in comparison with the experimental activity profiles for polymerizations carried out without prepolymerization from 50°C to 90°C and with 0.025 mol/l $H_2$. It can be seen that with the model the influence of reaction temperature on the activity profiles can be described in a quantitative good manner. Also the average activities calculated with the model are in a good agreement with the experimentally determined average activities (error between modeled and experimental data 4 - 17 %).

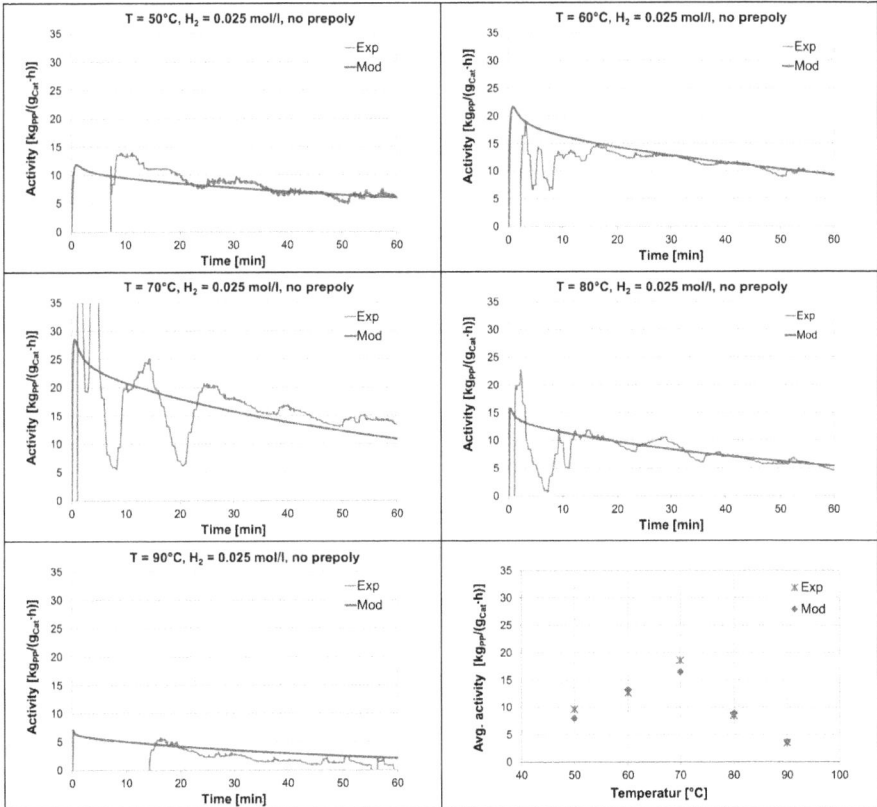

Figure 54: Results kinetic modeling catalyst B: Influence reaction temperature - Comparison simulated and experimental activity profiles and avg. activities (without prepolymerization, T=50-90°C, $H_2$=0.025 mol/l)

91

## 5.2.2.2 Influence of $H_2$ concentration

Figure 55 shows the comparison of simulated and experimentally derived activity profiles for the different hydrogen concentrations ranging from 0 to 0.05 mol/l exemplary for a reaction at 80°C. Calculated and experimentally determined average activities are within an error range between 4 - 14 %.

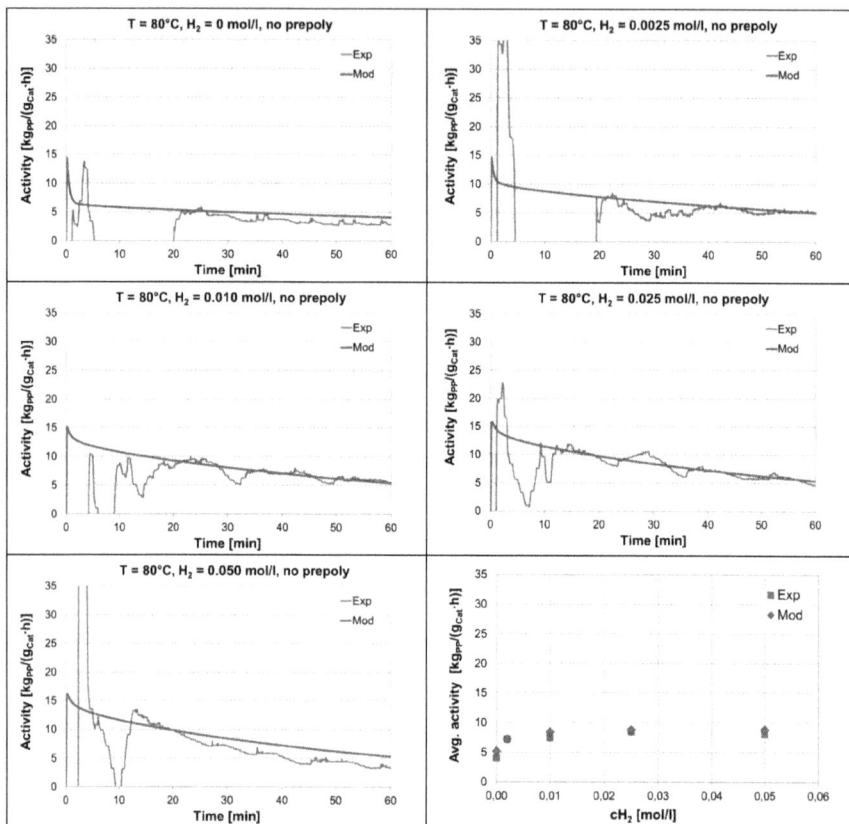

**Figure 55: Results kinetic modeling catalyst B: Influence $H_2$ concentration - Comparison simulated and experimentally derived activity profiles and avg. activities (without prepolymerization, T=80°C, $H_2$=0-0.05 mol/l)**

Figure 56 shows the calculated and experimentally derived weight average molecular weights. At 70°C, a good fit between modeled and experimental $M_W$ is achieved (error between 2 - 8 %). At higher temperatures, and especially for 0.01 - 0.025 mol/l $H_2$, the error between modeled and experimental $M_W$ is between 2 - 29 %.

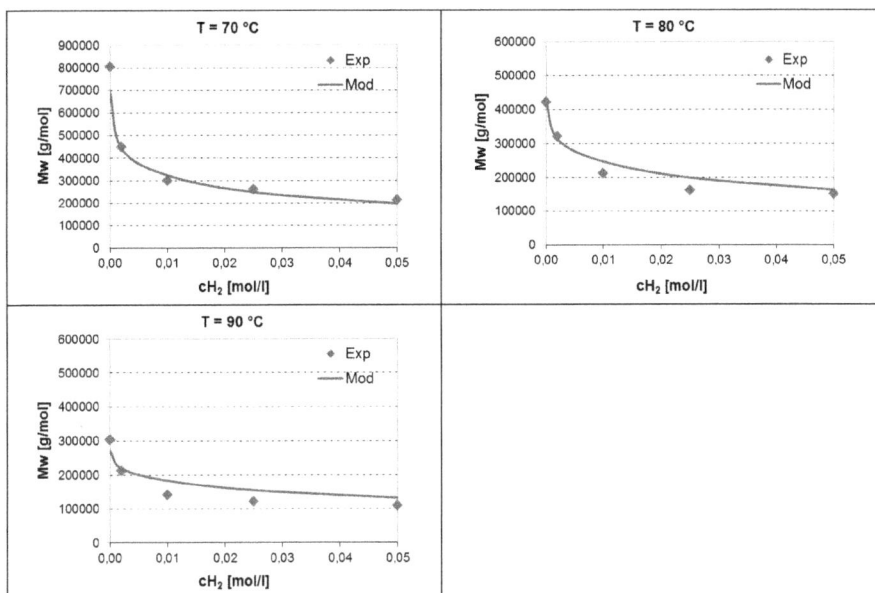

**Figure 56: Results kinetic modeling catalyst B: Influence H₂ concentration - Comparison simulated and experimentally derived avg. molecular weights (without prepolymerization, T=70-90°C, H₂=0-0.05 mol/l)**

### 5.2.2.3 Influence of injection conditions – effect of prepolymerization

In the model, the effect of the different catalyst injection conditions is described by the different growing factor g at the beginning of the polymerization at reaction temperature. In case of polymerization without prepolymerization step, the pure catalyst is injected directly at reaction conditions ($g^0$=1). The polymerization with prepolymerization step is shown after the prepolymerization step at main reaction conditions ($g^0$=18).

Figure 57 shows the simulated activity profiles as well as the corresponding particle temperature difference between particle temperature and gas-phase temperature for polymerizations with and without prepolymerization at 80°C. For polymerization without prepolymerization, the increase of the particle temperature at the beginning of the reaction is approx. three times higher compared to the particle temperature derived when a prepolymerized catalyst is used. Therefore, in the model, the thermal deactivation ($k_{des,temp}$) is more pronounced resulting in a much lower activity for polymerization without a prepolymerization step.

93

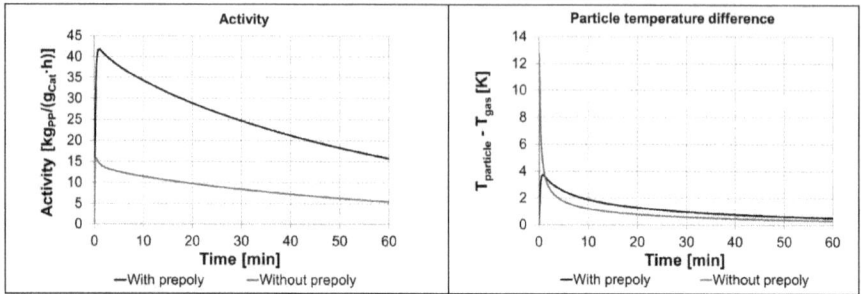

**Figure 57: Results kinetic modeling catalyst B: Influence of catalyst injection conditions left: on activity profiles, right: on particle temperature shown as temperature difference particle and gas-phase (T=80°C, $H_2$ = 0.025 mol/l)**

Table 18 shows the calculated particle temperature as well as the particle size for polymerizations with and without prepolymerization at 70°C in comparison with the experimental data. It can be seen that the calculated particle sizes are similar to the experimentally derived particle sizes with an error of less than 5 %.

**Table 18: Results kinetic modeling catalyst B: Influence of catalyst injection conditions: Comparison simulated particle temperature and particle size (T=70°C, $H_2$ = 0.025 mol/l)**

| prepolymerization | Model | | | | Experiment | | Error |
| --- | --- | --- | --- | --- | --- | --- | --- |
| | $T_P^0$ | $\Delta T_p$ | $D_P^0$ | $D_P^{60}$ | $D_{50}^0$ (Malvern) | $D_{50}^{60}$ (Sieving) | $D_P^{60}$ |
| | °C | K | µm | µm | µm | µm | % |
| No ($g^0$=1) | 84.5 | 14.5 | 55 | 1822 | 55 | 1773 | 2.7 |
| Yes ($g^0$=18) | 73.5 | 3.5 | 990 | 2187 | 929 | 2117 | 3.2 |

The comparison of simulated and experimentally determined average activities for both procedures is shown in Figure 58. For polymerizations without prepolymerization, calculated and experimental avg. activities are over the whole temperature range within an error of 4 - 17 %. For polymerizations with prepolymerization step, the trends of the average activities are within an error of 3 - 17 %.

94

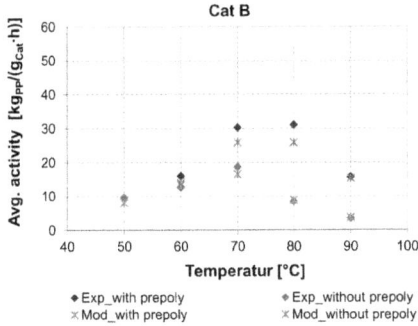

**Figure 58: Results kinetic modeling catalyst B: Influence of catalyst injection conditions – Comparison simulated and experimentally derived avg. activities with and without prepolymerization ($H_2$ = 0.025 mol/l)**

The comparison of the simulated and experimentally derived activity profiles for polymerizations with prepolymerization from 60°C to 90°C is shown in Figure 59. While at 70°C and 80°C the activity profiles are slightly underestimated, it can be seen that the activity profile at 90°C is slightly overestimated. The deviations can be attributed maybe to the same growing factor g used at the different reaction temperatures.

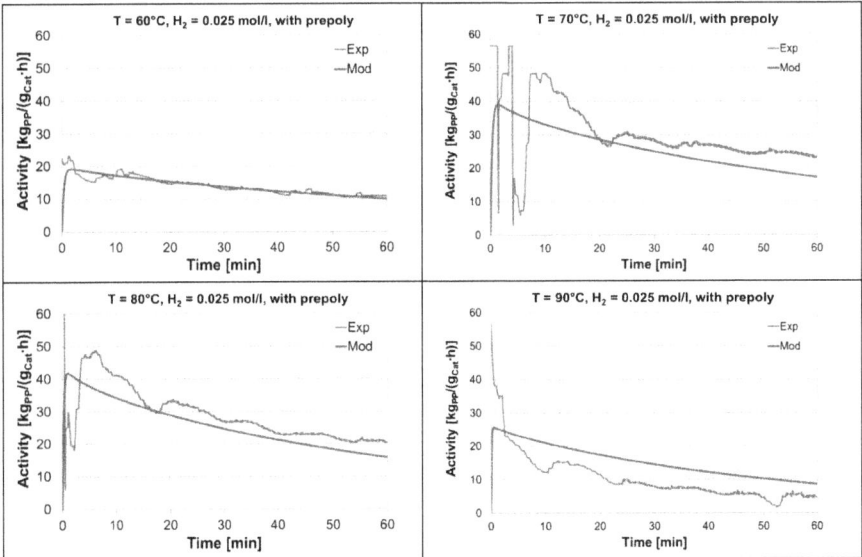

**Figure 59: Results kinetic modeling catalyst B: Influence of catalyst injection conditions – Comparison simulated and experimentally derived activity profiles (with prepolymerization, T=60-90°C, $H_2$ = 0.025 mol/l)**

95

### 5.2.3 Comparison of simulated and experimental data – ZN catalyst A

In this chapter, only selected results for catalyst A are shown as similar tendencies as described for catalyst B are obtained. As described earlier, the same kinetic model with the same estimated kinetic parameters was used for the simulation of the polymerization. The differences of ZN catalyst A are only described with the different amount of active component.

The comparison of simulated and experimentally derived activity profiles for polymerizations with prepolymerization is shown in Figure 60. For polymerizations with prepolymerization, avg. activities are calculated quite well with an error of approx. 11 % ± 4 %. For polymerizations without prepolymerization step, calculated and experimentally derived avg. activities at low temperatures are in a good agreement. For higher reaction temperatures of 70°C and 80°C, avg. activities are slightly underestimated with an error of 15 - 20 %. At 90°C, the error seems with 40 % high but absolute values of the avg. activities are quite similar (experiment: 5.2 $kg_{PP}/g_{Cat}/h$ model: 2.9 $kg_{PP}/g_{Cat}/h$).

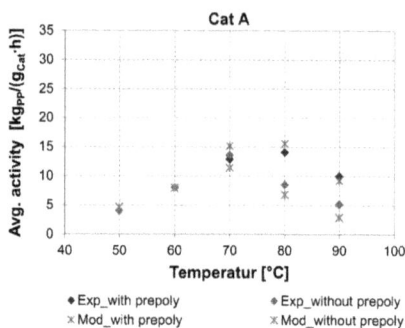

Figure 60: Results kinetic modeling catalyst A: Influence of catalyst injection conditions – Comparison simulated and experimentally derived avg. activities with and without prepolymerization ($H_2$ = 0.025 mol/l)

Figure 61 shows the comparison of simulated and experimentally derived activity profiles in dependence of the different reaction conditions (T, $H_2$, prepolymerization). It can be seen that the course of the activity profile can be predicted for catalyst A with the same kinetic model and set of kinetic constants. The influence of different reaction temperatures, $H_2$ concentrations as well as catalyst injection conditions on the activity profile can be simulated in a quantitative good manner. Only at higher reaction temperatures of 80°C and 90°C the beginning is underestimated, which can be related to the strong influence of the thermal deactivation at the beginning of the reaction at these high temperatures.

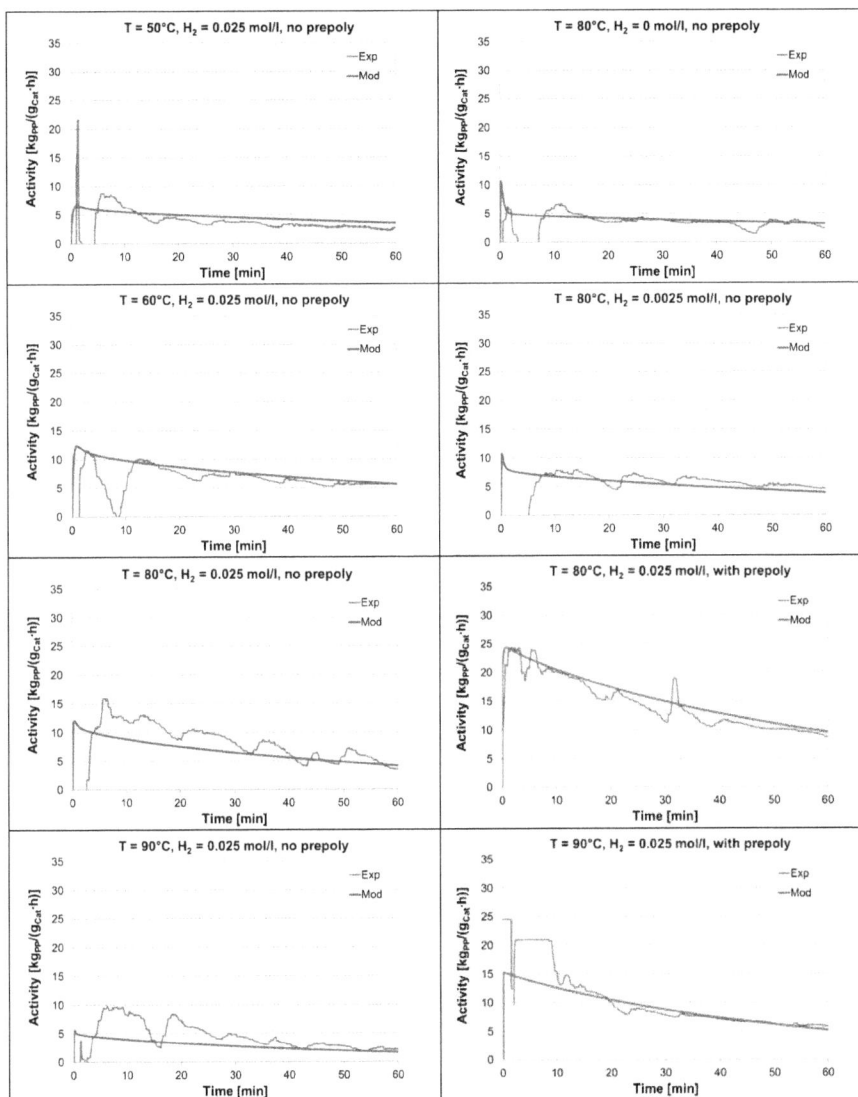

**Figure 61: Results kinetic modeling catalyst A: Influence of different reaction conditions – Comparison simulated and experimentally derived activity profiles (with and without prepolymerization, T=50-90°C, $H_2$ =0-0.025 mol/l)**

The comparison of calculated and experimentally estimated weight avg. molecular weights is shown in Figure 62. The error between modeled and experimentally derived $M_W$ is in the range of 0.6 - 16.6 %.

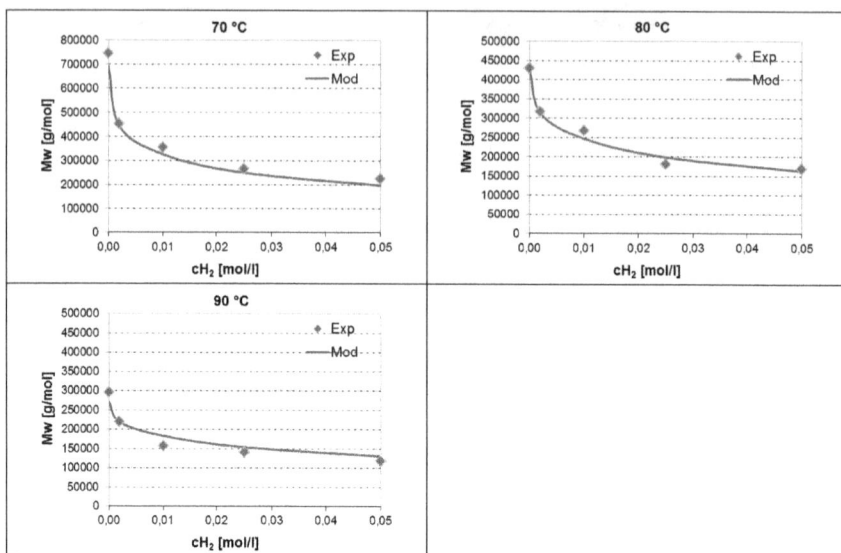

Figure 62: Results kinetic modeling catalyst A: Influence $H_2$ concentration - Comparison simulated and experimentally derived avg. molecular weights (without prepolymerization, T=70-90°C, $H_2$=0-0.05 mol/l)

## 5.3 Summary kinetic modeling of the gas-phase polymerization of propylene with Ziegler-Natta catalysts

Aim of the modeling was to develop a simplified phenomenological kinetic model for the gas-phase polymerization of propylene with two Ziegler-Natta catalysts which can be used to describe the influence of different reaction conditions, including the effect of prepolymerization, on activity as well as to calculate the average molecular weights.

The model is based on a quasi-homogeneous particle model neglecting mass-transfer and considering only one sort of active sites. The kinetic scheme was derived based on the experimentally observed activity-time profiles and weight average molecular weights. As elementary steps formation of active sites, chain propagation, chain transfer (by $H_2$ and spontaneous) as well as spontaneous deactivation were considered. The influence of $H_2$ on the activity was described based on the dormant site theory. Monomer concentrations at the active sites were calculated with Henry's law and Stern equation. Furthermore, the $H_2$ concentration was assumed to be equal to the gas-phase concentration. Averages of the molecular weight distribution were calculated by the method of moments. Based on the hypothesis that particle overheating is the main reason for the strong decrease in activity observed on direct injection of catalysts, the effect of injection conditions was modeled by consideration of the particle growth, calculation of the particle heat balance and consideration of an additional, highly temperature dependent thermal deactivation reaction. For heat removal, the Ranz-Marshall correlation was used. The temperature dependency of the kinetic constants was calculated with the Arrhenius equation. Therein, temperatures are based on the particle temperature. The outlined model was implemented in the simulation package gPROMS.

As in the experiments same trends of the activities as well as same hydrogen response were observed for both catalysts, the same kinetic scheme was used for modeling. Model parameters were determined stepwise by parameter estimation to the experimental results. First estimations showed that similar kinetic parameters are derived for both catalysts. Thus, same final kinetic parameters were estimated for both catalysts. The only differentiating parameter between both catalysts is the fraction of active sites. Comparison of simulation results and experimental data reveal that the derived model can be used to predict the average activities, activity profiles as well as the molecular weights for both ZN catalysts at the different reaction conditions. Also the effect of prepolymerization resp. initial particle overheating at higher reaction temperatures on catalyst activity can be described in a quantitative good manner.

# 6 Experimental investigation of the bulk polymerization of propylene with a supported metallocene catalyst

## 6.1 Scope and outline

Scope of the second part of the work is the kinetic investigation of a supported metallocene catalyst for the polymerization of propylene under bulk conditions and the development of a kinetic model describing the polymerization.

For the measurement of the reaction kinetics of the bulk polymerization, a reaction calorimeter was used which works based on the heat flow calorimetry (see chapter 2.5.2). Therein, an experimental setup has been built up around an existing reaction calorimeter and appropriate experimental procedures for reliable and systematic kinetic measurements have been developed. In chapter 6, the experimental study on the influence of different reaction conditions (reaction temperature, hydrogen concentration) on reaction kinetics and resulting polymer characteristics is reported. A particular focus is set – as in the previous chapters – on the investigation of the influence of catalyst injection conditions (prepolymerization) on polymerization behavior of the catalyst system.

Based on the obtained experimental data, in chapter 7, a simplified kinetic model for the description of the bulk polymerization with metallocene catalyst is developed and the kinetic parameters of the catalyst are determined.

## 6.2 Experimental setup for the bulk polymerizations

For the experimental investigation of the bulk polymerization of propylene with a supported metallocene catalyst, a new laboratory setup was build up within this work. Main part of the setup is a special reaction calorimeter from ChemiSens®, which also provides a new method for the online-measurement of the kinetics of the propylene bulk polymerization. A scheme of the setup is shown in Figure 63. It consists of the following sections:

- raw material supply and improved purification system for liquid propylene
- propylene and hydrogen feeding system
- reaction calorimeter.

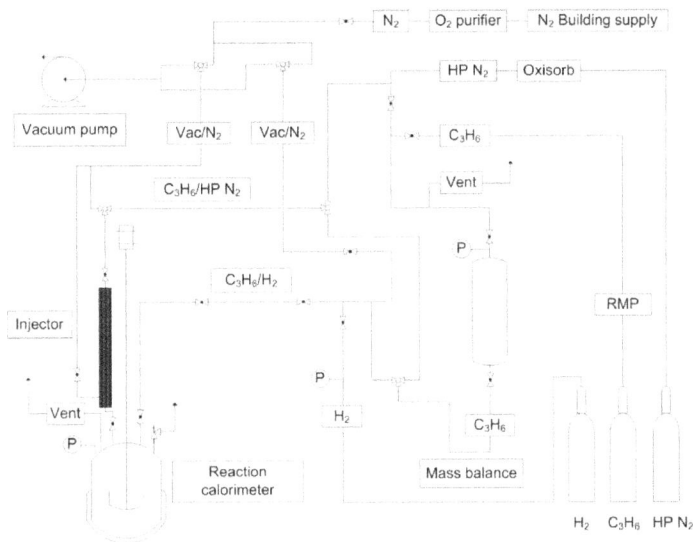

Figure 63: Scheme experimental setup for the bulk phase polymerization

## 6.2.1 Improved raw material purification

Metallocene catalysts are much more sensible to impurities compared to Ziegler-Natta catalysts. In a set of pre-studies it was found out that the purity of the monomer was not high enough to ensure the desired catalyst activity. Therefore, the existing raw material purification system for propylene was further improved. In the discussion with Dr. Karrer (BASF), a modified six-stage propylene purification system was developed, which is specified for cleaning liquid propylene feed streams. Therein, the two purification catalysts of PuriStar® R3-16 were replaced by the catalysts PuriStar® R3-12 and PuriStar® R3-17 in reduced form (both from BASF). The PuriStar® R3-12 is made of CuO/ZnO tablets and is used to remove arsine, phosphine and reactive sulfur from the feed stream. Commercially it is widely used in the purification of refinery, chemical and polymer grades of propylene. The PuriStar® R3-17 consists of CuO tablets and is used to remove traces of CO and $O_2$ especially from liquid propylene. For the removal of $H_2O$ also molecular sieve of 3-4 Å was used. In addition, a mixture of the adsorbents Selexsorb CD and Selexsorb COS (BASF) was used to remove $CO_2$ and last traces of water. Figure 64 shows schematically the revised six-stage purification system.

101

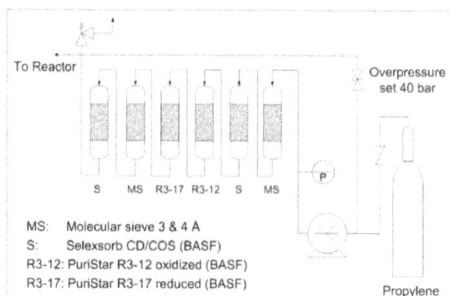

**Figure 64: Improved raw material purification for liquid propylene**

## 6.2.2  Propylene and hydrogen feeding

In order to feed the desired amount of propylene into the reactor and to avoid overfilling, a steel cylinder placed on a weight balance was used. Herein, the desired amount of propylene was weighted out and flushed into the reactor. In order to ensure a flow from the cylinder to the reactor, the reactor temperature was kept low (10 - 15°C) while the measuring cylinder was heated up. Due to the increased vapor pressure above the liquid propylene in the heated measuring cylinder, the propylene is forced to flow into the reactor.

Hydrogen was added batch-wise using a defined tube volume and known partial pressure. From the hydrogen gas bottle a back-pressure of 34 bar was adjusted. The tube was completely filled with hydrogen and flushed into the evacuated reactor. This procedure was repeated until the desired amount of hydrogen was fed into the reactor.

## 6.2.3  Reaction calorimeter

The bulk polymerizations were carried out in the reaction calorimeter CPA 202 (ChemiSens®). The cylindrical reactor is made of hastelloy and has a volume of 250 ml. The reactor is certified for temperatures from -50 °C up to 200°C and pressures from vacuum up to 100 bars. The reactor is equipped with a magnetic coupled anchor stirrer whereby stirring speeds from 50 to 2000 rpm are possible. The reactor is placed in a thermostating unit filled with water as thermostating liquid. Several transducers are connected to the reactor and the thermostating unit for measurement of process variables such as temperatures, pressure, stirrer torque and speed as well as heat flow. All transducers are connected to the control unit which communicates with a RS232 connection with the computer. The calorimeter system is operated with the proprietary software ChemiCall V2 (ChemiSens®) which records and evaluates the measured data and calculates further process data such as power input of the stirrer, thermal power and heat transfer coefficient. A picture of the whole setup and the reactor is shown in Figure 65.

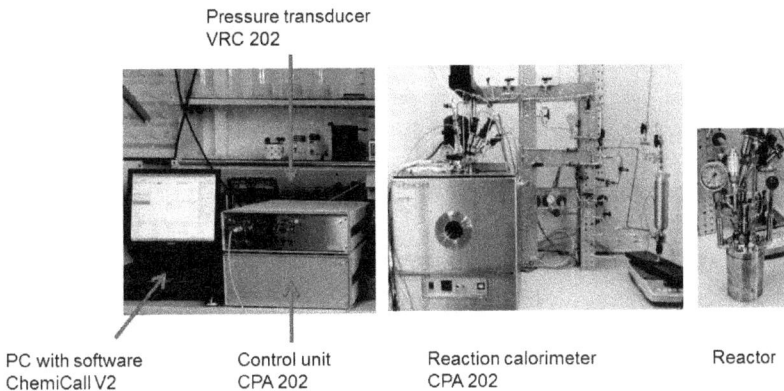

Figure 65: Setup reaction calorimeter ChemiSens® CPA 202

### 6.2.4 Measuring principle

The used reaction calorimeter works according a unique, calibration-free heat flow principle, wherein the heat flow generated by the exothermal polymerization reaction is measured by heat conductivity in the reactor base [130]. A scheme of the measuring principle of the reaction calorimeter is shown in Figure 66.

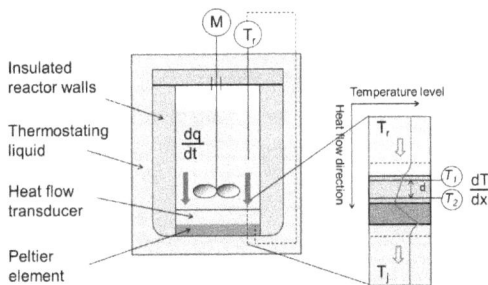

Figure 66: Measuring and heat pumping principle of reaction calorimeter CPA202 ChemiSens® [130], [144]

The calorimeter works in isothermal mode keeping the reactor temperature and bath temperature constant during reaction. The walls of the reactor are insulated. In order to avoid any internal heat flux, the temperature of the surrounding thermostating liquid is always kept 0.2°C higher than the reactor temperature (active insulation). The heat flow can only pass through the reactor base which consists of a steel plate and a Peltier element. The Peltier element acts as reversible heat pump between the reactor and the surrounding thermostating liquid (heat sink) and has a heating/cooling capacity of 30 W. It generates the necessary temperature difference in order to drive the heat flow out of the reactor. Temperature sensors (heat flow transducer), which are placed in a defined distance between reactor bottom and Peltier element, are measuring the temperature difference. The heat flow generated by the polymerization reaction is then easily calculated by means of the heat conductivity equation in the reactor base:

$$\dot{Q}_{HF,reactor\ base} = \lambda \cdot \frac{A}{d}(T_1 - T_2) \qquad [W]$$
(6.1)

where $\lambda$ is the specific heat conductivity coefficient of steel (W/(m·K)), $A$ is the defined area of the reactor base (m²), $d$ is the distance between the temperature sensors (m) and $T_1$- $T_2$ is the measured temperature difference in the reactor base (K).

The total heat balance of the reaction calorimeter is shown in equation (6.2). Therein, the chemical heat produced by the polymerization ($\dot{Q}_{chem}$) is equal to the measured heat flow through the reactor base ($\dot{Q}_{HF,reactor\ base}$) minus the stirring power ($P_{Stirr}$) introduced by the stirrer plus the accumulated heat in the reactor ($\dot{Q}_{accumulation}$). Heat losses to the surrounding thermostating bath can be neglected due to the active insulation [144].

$$\dot{Q}_{chem} = \dot{Q}_{HF,reactor\ base} - P_{Stirr} + \dot{Q}_{accumulation} \qquad [W]$$
(6.2)

Once the reactor operates at isothermal conditions, the accumulation term disappears and the heat balance simplifies to:

$$\dot{Q}_{chem} = \dot{Q}_{HF,reactor\ base} - P_{Stirr} \qquad [W]$$
(6.3)

The main advantage of the calorimeter to other calorimeters is that measurement of the heat flow is independent of the filling level and heat transfer conditions in the reactor. Hence, calibrations for determination of heat transfer coefficients are not necessary.


## 6.3   Chemicals

Liquid propylene (purity 2.5, Linde, Air Liquide) was used as monomer which was purified in the six-stage raw material purification system (see chapter 6.2.1). Hydrogen (purity 6.0, Linde) was used as chain transfer agent without further purification. For the injection of the catalyst, high pressure nitrogen (purity 5.0, Linde) was used which was purified with an oxisorb-cardrige (Linde) in order to remove possible solved oxygen. Nitrogen from the building supply (high grade, Air Liquide) was used to clean and inertize the system. The bulk polymerization was carried out with a supported metallocene catalyst provided by the industrial cooperation partner.

The delivered catalyst was already activated and no additional cocatalyst was needed. As scavenger a 1.0 M solution of triisobutylaluminum (TIBA, Sigma-Aldrich) was used to remove last impurities in the reactor.

## 6.4 Polymerization procedures

Polymerization experiments with the high active metallocene catalyst were always carried out with a prepolymerization step in order to avoid particle overheating as described in chapter 2.4. Two different polymerization procedures were developed: In the first procedure, the polymerization was carried out with an in-situ prepolymerization. In the second procedure, the polymerization was carried out with a prepolymerized catalyst which was prepared in a separate step (external prepolymerization). Aim was to define a suitable polymerization procedure with which accurate and reproducible kinetic measurements with a fast access to the kinetic data can be carried out. Both procedures were compared and a final polymerization procedure was chosen to carry out the kinetic measurements.

### 6.4.1 General reaction preparations

Because metallocene catalysts are very sensible to any impurities, an intensive cleaning and inertization of the reaction calorimeter and the tubing system was always necessary. For the inertization, a two-stage rotary vane vacuum pump (P 6 Z, Ilmvac) and nitrogen from the building supply was used. The reactor was heated up to 80°C. The reactor as well as the whole tubing system was evacuated for 30 minutes and then flushed with nitrogen. This cycle was repeated at least six times. At the end, fast change of vacuum and nitrogen was carried out several times. The reactor was set under nitrogen overpressure and cooled down. Shortly before the experiment was started, the reactor was flushed three times with fresh propylene and evacuated.

The catalyst and scavenger were prepared in the glove-box (Jacomex) under inert conditions ($N_2$ atmosphere) in separate feeders in order to avoid any pre-contact. The feeders are similar to the construction shown in Figure 15, page 35, but with only one chamber. TIBA, which was used as scavenger, was injected with a syringe into the first feeder. The weighted amount of catalyst was either added dry or together with 1 ml heptane into the second feeder. Both feeders were connected successively to the system (according to the polymerization procedure) under nitrogen flow. The connections were inertized with alternating vacuum and $N_2$ flush.

## 6.4.2 Polymerization with in-situ prepolymerization

A scheme of the first polymerization procedure is shown in Figure 67. Hydrogen, propylene and scavenger were added into the inertized and evacuated reactor. Therein, half of the propylene feed was used to flush the $H_2$ feeding line, while the other half of the propylene was used to flush the scavenger into the reactor. The reactor was then heated up to 40°C giving the scavenger time to react with impurities in the monomer and in the reactor. The catalyst was injected using high pressure nitrogen. The prepolymerization was carried out for approx. 15 min while the reactor was heated up to the desired reaction temperature. The main polymerization was then carried out for one hour. The reaction was stopped by flushing off the liquid propylene which simultaneously led to a rapid decrease of the reaction temperature. The catalyst was finally deactivated by contact with compressed air.

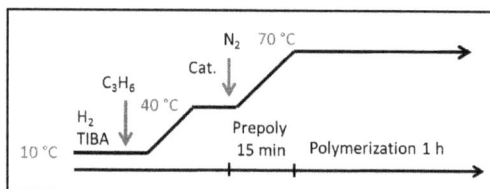

Figure 67: Bulk polymerization procedure A – Polymerization with in-situ prepolymerization

## 6.4.3 Polymerization with external prepolymerization

In the in-situ prepolymerization outlined in the previous paragraph, the process is non-isothermal as the reactor is heated-up during prepolymerization to the main polymerization temperature. Hence, conditions are changing and are less defined. Furthermore, the exact degree of prepolymerization is not known and, in fact, the transition between prepolymerization and main polymerization stage is not sharp but continuous. Since these uncertainties limit the possibilities for optimization, a further prepolymerization procedure was developed.

In the second polymerization procedure, a separate batch of prepolymerized catalyst was produced firstly at mild reaction conditions. The prepolymer was then isolated and stored in the glove-box under inert conditions. In the second step (main polymerization), the prepolymerized catalyst was injected directly at reaction conditions.

Production of the prepolymerized catalyst

The reactor was cleaned and inertized as described previously. Hydrogen, propylene and scavenger were added. A larger amount of dry catalyst (20 to 25 mg) was prepared in the glove-box and injected with liquid propylene into the reactor. The polymerization was carried out at 25°C for a defined time until a desired degree of prepolymerization (DP) was reached. The reaction was interrupted by releasing the monomer and flushing with nitrogen while cooling down the reactor. The reactor was further flushed with $N_2$ in order to remove all of the remaining propylene. The pressurized reactor (with nitrogen) was then dried for several hours in an oven in order to remove any water from the outer reactor surface. Afterwards, the reactor was inserted into the glove-box, where the prepolymerized catalyst was removed and weighted out under inert conditions. With this procedure it can be ensured that the catalyst remains active. The exact degree of prepolymerization is calculated from the weighed amount of produced prepolymer per initial amount of catalyst according following equation:

$$DP = \frac{m_{PP}}{m_{Cat}} \qquad \left[\frac{mgPP}{mgCat}\right] \tag{6.4}$$

Table 19: Reaction conditions for the production of prepolymerized catalyst

| Conditions | Amount | Notice |
|---|---|---|
| Liquid propylene | 75 g | Approx. 150 ml at 23°C (NIST[132]) |
| Catalyst MC-1 | 20 – 25 mg | Dry |
| Hydrogen | 2.0 mg | For max. activity |
| TIBA | 0.8 mmol | 1M solution of TIBA in hexane |
| Reaction temperature | 25 °C | |
| DP | 50 – 200 mg$_{PP}$/mg$_{Cat}$ | Corresponding $t_{poly}$ = 15 – 60 min |

Main polymerization

Figure 68 shows schematically the second polymerization procedure for the main polymerization carried out with prepolymerized catalyst. Hydrogen and scavenger were added together with propylene into the inertized reactor which was then heated up to the desired reaction temperature. When a constant reaction temperature was reached, the dry prepolymerized catalyst was injected with high pressure $N_2$. The polymerization as well as the activity measurements directly started with the injection of the catalyst. The reaction was carried out for one hour and was stopped by flushing off the monomer. The catalyst was also deactivated by using compressed air.

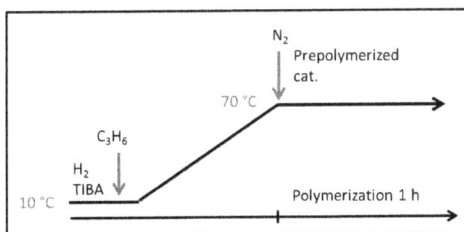

Figure 68: Bulk polymerization procedure B – Polymerization with prepolymerized catalyst

## 6.5 Reaction conditions and experimental plan

The reaction conditions for the main polymerization of both procedures are summarized in Table 20. The reaction conditions for the production of prepolymerized catalyst were already shown in Table 19. The calorimetric measurements were carried out in isothermal mode. Stirring speed was kept always at 350 rpm ensuring a proper mixing of the reaction medium and good heat transfer conditions. A torque transducer was used to measure the heat input of the stirrer. In order to avoid overfilling of the reactor and for safety reasons, the reactor was filled with 75 g propylene (approx. 150 ml at 23 °C, NIST database [132]). In several pre-studies it was found that a scavenger amount of 0.8 mmol (1M solution of TIBA in hexane) is necessary to ensure the desired catalyst activity. Polymerizations with procedure A were carried out with 1 mg catalyst dispersed in 1 ml heptane. This was necessary for catalyst preparation and to ensure good injection into the reactor. In procedure B, the used amount of prepolymerized catalyst with a defined DP was adjusted having a total amount of 1 mg of pure catalyst for the polymerization reaction. Due to the larger particle size of the prepolymerized catalyst, no heptane was needed for the preparation and injection.

Table 20: Reaction conditions for the bulk polymerization of propylene with metallocene catalyst

| Component | Amount | Notice |
|---|---|---|
| Liquid propylene | 75 g | Approx. 150 ml at 23°C (NIST [132]) |
| Catalyst | 1 mg | Proc. A: dispersed in 1 ml heptane<br>Proc. B: dry, calculated according DP |
| Hydrogen | 0 to 5 mg | According experimental plan |
| TIBA | 0.8 mmol | 1M solution of TIBA in hexane |
| Polym. temperature | 50 – 80 °C | According experimental plan |
| Stirrer speed | 350 rpm | |

The kinetic study of the metallocene catalyst for the polymerization in liquid propylene was carried out in the following operational window:

- $T = 55 - 80°C$
- $H_{2,feed} = 0 - 5$ mg ($0 - 0.14$ mol%)

Furthermore, the influence of the prepolymerization degree on the catalyst activity was tested. Therefore, prepolymerized catalysts with a DP ranging from 50 to 200 $mg_{PP}/mg_{Cat}$ were prepared and investigated in a standard polymerization at 70°C and with 2 mg $H_2$ ($c_{H2,feed} = 0.055$ mol%). Due to the high sensitivity of the catalysts and in order to test the reproducibility, the measurements were carried out several times. The detailed experimental plan of the measurements is shown in Table 21.

Table 21: Experimental plan for bulk polymerization with metallocene catalyst

| $H_{2,feed}$ in mg<br>$H_{2,feed}$ in mol% | 0<br>0 | 1<br>0.027 | 2<br>0.055 | 4<br>0.110 | 5<br>0.140 |
|---|---|---|---|---|---|
| $T_{Poly}$ in °C | | | | | |
| 55 | V90 | V103 | V86 | V91 | |
| 65 | V95 | V107, V109 | V94 | V96 | |
| 70 | V88 | V97, V108 | V87, V99 | V89 | V98 |
| 80 | V101 | V104, V110 | V100 | V102 | |
| 70 | | | DP 50-200 | | |

## 6.6 Analytics

The produced polymer samples were analyzed using analytical methods already described in chapter 4.5. MFR, molecular weights and molecular weight distributions as well as crystallinity of the polymer samples were measured by the industrial cooperation partner. Additionally, MFR of selected polymer samples were measured in the own laboratory using a micro flow melt indexer (CSI-127MF, Custom Scientific Instruments, Inc.) in order to build up a correlation between measured MFR and $M_w$ from GPC measurements (test method and conditions see chapter 4.5.1). Porosity, bulk density as well SEM measurements were carried out at the Martin-Luther-University Halle-Wittenberg.

## 6.7 Gained experimental data and calculation of activities

### 6.7.1 Heat flow measurements

Figure 69 shows the gained experimental data for a polymerization reaction at 70°C according to procedure B. The shown heat flow curve over reaction time corresponds to the total measured heat flow through the reactor base, where heat accumulation is already considered. At constant reaction temperature, the measured heat flow corresponds to the sum of energy released by the polymerization reaction and power input by the stirring system (see chapter 6.2.4). The power input of the stirrer system is determined independently by the measurement of torque and stirrer speed.

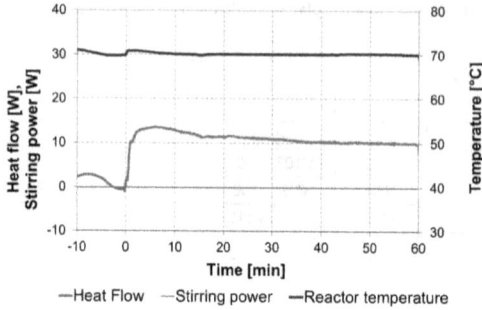

Figure 69: Gained experimental data from calorimetric measurements (example polymerization at 70°C according procedure B)

As the example in Figure 69 shows, the stirring power is nearly zero and remains constant during reaction time. This is an expected result as for polymer dispersions the increase in viscosity with the polymer content is much less as for polymer solutions. Nevertheless, power input of the stirring system is always considered for the calculation of the activity described in the following.

## 6.7.2 Calculation of activity

At isothermal conditions, the chemical heat produced by the polymerization reaction can be determined from the measured total heat flow through the reactor base reduced by the measured heat input of the stirrer (see equation (6.3), chapter 6.2.4). It is equal to the reaction volume $V_R$, the overall rate of polymerization $R_P$ and the reaction enthalpy $(-\Delta_R H)$:

$$\dot{Q}_{chem} = \dot{Q}_{HF} - P_{Stirr} = V_R \cdot R_P \cdot (-\Delta_R H) \qquad [W] \qquad (6.5)$$

The current activity of the catalyst, which is defined as amount of produced polymer per amount of catalyst and time, can be then calculated from the heat flow measurements according following equation:

$$A = \frac{\dot{m}_{pol}}{m_{Cat}} = \frac{V_R \cdot R_P \cdot MW}{m_{Cat}} = \frac{\dot{Q}_{chem} \cdot MW}{m_{Cat} \cdot (-\Delta_R H)} \cdot 3600 \frac{s}{h} \qquad \left[\frac{kg_{Pol}}{g_{Cat} \cdot h}\right] \qquad (6.6)$$

The average activity can be determined experimentally by weighting out the amount of produced polypropylene related to the amount of catalyst for a reaction time of 1 hour:

$$\bar{A} = \frac{m_{pol,total}}{m_{Cat} \cdot h} \qquad \left[\frac{kg_{PP}}{g_{Cat} \cdot h}\right] \qquad (6.7)$$

It can also be determined from the calorimetric measurements. Therein, the total amount of produced polymer is calculated by integrating the heat flow curve over reaction time:

110

$$m_{pol,total} = \frac{\int_0^t \dot{Q}_{chem} dt \cdot MW}{(-\Delta_R H)} \qquad [kg_{Pol}] \qquad\qquad (6.8)$$

The reaction enthalpy used in equations (6.6) and (6.8) for the polymerization of liquid propylene to solid polypropylene is $\Delta_R H$ = -83 kJ/mol [137].

### 6.7.3 Validation calorimetric measurements

In order to validate the calorimetric measurements, the calculated avg. activities are compared with the avg. activities derived from the yield and are shown in the parity diagram in Figure 70.

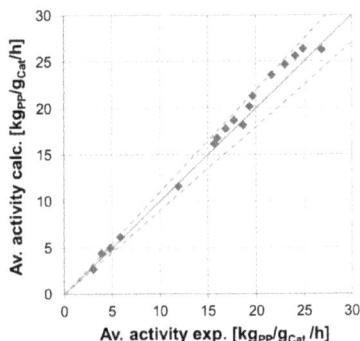

Figure 70: Validation calorimetric measurements

The calculated avg. activities from the calorimetric measurements are in a good agreement with the experimental avg. activities within a 10 % error range. The deviations could be related to measuring errors of the experimental avg. activities or to the correct starting point for the integration of the heat flow curve. Thus, the calorimetric measurements can be used to calculate the activities and to determine the kinetics of the polymerization reaction.

### 6.7.4 Reproducibility

The reproducibility of both polymerization procedures was tested at standard polymerization conditions (70°C, $c_{H2,feed}$ = 0.055 mol%). In procedure B, a prepolymerized catalyst with a DP of 100 $mg_{PP}/mg_{Cat}$ was used which was always injected at 40°C. The resulting avg. activities as well as the MFR values are summarized in Table 22. For both polymerization procedures similar avg. activities as well as MFR values are obtained.

Table 22: Reproducibility of bulk polymerization: Avg. activity and MFR values for procedure A and B
(T=70°C, prepolymer: DP=100 mg$_{PP}$/mg$_{cat}$)

| Procedure | Exp. | m$_{cat}$ [mg] | TIBA [mmol] | m$_{H2}$ [mg] | Activity [kg$_{PP}$/g$_{cat}$/h] | MFR [g/10min] |
|---|---|---|---|---|---|---|
| A | 28 | 1.12 | 0.8 | 1.93 | 20.2 | 8.6 |
| A | 29 | 1.12 | 0.8 | 2.04 | 20.4 | 8.3 |
| A | 31 | 1.12 | 0.8 | 1.97 | 19.6 | 9.3 |
| B | 49 | 0.900 | 0.8 | 1.78 | 19.3 | 10.5 |
| B | 50 | 0.922 | 0.8 | 1.92 | 20.1 | 11.3 |
| B | 51 | 0.876 | 0.8 | 1.91 | 20.0 | 10.1 |

Figure 71 shows as an example for procedure B the activity profiles calculated from the measured heat flow according to equations (6.5) to (6.7).

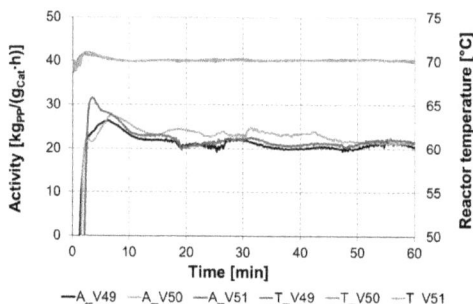

Figure 71: Reproducibility of bulk polymerization: Activity profiles (procedure B, T=70°C, c$_{H2,feed}$=0.055 mol%, prepolymer: DP=100 mg$_{PP}$/mg$_{cat}$)

## 6.8 Results kinetic measurements of the bulk polymerization with a supported metallocene catalyst

### 6.8.1 Comparison of both polymerization procedures

In order to compare both procedures, polymerizations were carried out at the same standard reaction conditions (70°C, c$_{H2,feed}$ = 0.055 mol%). The resulting average activities and the corresponding MFR values are shown in Figure 72.

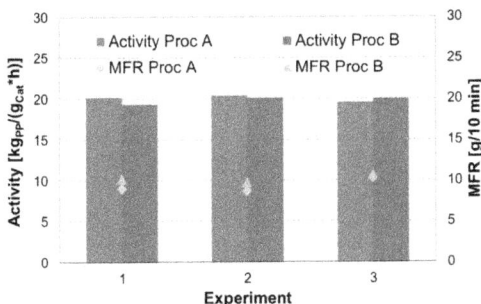

Figure 72: Results bulk polymerization: Comparison of polymerization procedures A) with in-situ prepolymerization and B) with external prepolymerization - Avg. activity and MFR value ($T_{poly}$=70 °C, $c_{H2,feed}$=0.055 mol%, $DP_{prepolymer}$=100 $mg_{PP}/mg_{Cat}$)

As it can be seen, the reached average activities as well as the MFR values of the produced polymer are similar for both polymerization procedures. The results could also be reproduced in all three experiments.

The differences of both procedures are visible in the measured heat flow curves shown in Figure 73. As described in chapter 6.7.2, kinetic information are only accessible at isothermal conditions, where the measured heat flow is proportional to the current activity of the catalyst.

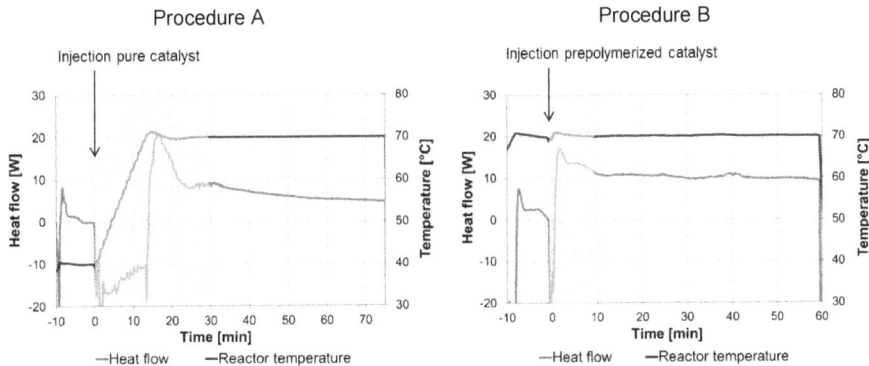

Figure 73: Results bulk polymerization: Comparison of polymerization procedures A) with in-situ prepolymerization and B) with external prepolymerization – Measured heat flow curves ($T_{poly}$=70°C, $c_{H2,feed}$=0.055 mol%, $DP_{prepoylmer}$=100 $mg_{PP}/mg_{Cat}$)

In case of procedure A, there is a big time lag of approx. 30 min until isothermal conditions were reached. During this time no kinetic information can be derived from the heat flow measurements. In procedure B, the prepolymerized catalyst is directly injected at reaction conditions. After a small temperature increase, isothermal conditions were reached within 10 min. The kinetic information are therefore earlier accessible in comparison to procedure A. Another advantage of procedure B is that the prepolymerized catalyst is produced at defined

113

reaction conditions. There is no undefined heating-up period during the prepolymerization as in procedure A and the DP can be exactly calculated from the derived prepolymer.

Based on the comparison of both procedures, it was therefore decided to carry out the final kinetic measurements according procedure B with prepolymerized catalysts.

## 6.8.2 Influence of the degree of prepolymerization on the catalyst activity

Prepolymerized catalysts with different prepolymerization degrees from 50 to 200 $mg_{PP}/mg_{Cat}$ were prepared and compared in a main polymerization in order to investigate its influence on the resulting avg. activity of the catalyst. The main polymerizations were carried out under standard polymerization conditions (T = 70°C, $c_{H2,feed}$ = 0.055 mol%) and were also repeated several times testing the reproducibility of the results. The derived avg. activities for the different prepolymerization degrees are shown in Figure 74.

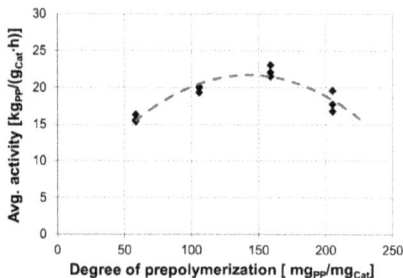

Figure 74: Influence of the prepolymerization degree on the average activity (T=70°C, $c_{H2,feed}$=0.055 mol%)

It can be seen that a maximum avg. activity was reached with the prepolymerized catalyst with a DP of 160 $mg_{PP}/mg_{Cat}$. For the polymerizations carried out with prepolymerized catalyst with lower DP or higher DP, respectively, lower activities were reached.

A possible explanation for the lower activities derived with prepolymer with lower DPs of 50 and 100 $mg_{PP}/mg_{Cat}$ might be related to the smaller prepolymer particles having lower surface area compared to prepolymers with higher DP. With the injection of the prepolymer at reaction conditions, high temperatures and high monomer concentration lead to high reaction rates. The particle temperature would increase more due to the lower heat transfer areas which maybe lead to higher particle temperatures (particle overheating) influencing the activity of the catalyst.

This could not be the explanation for prepolymers with a higher DP of 200 $mg_{PP}/mg_{Cat}$. This effect is not clear yet. The decrease of the activity might be due to possible transfer limitations of the monomer to the active sites of the prepolymerized catalyst. But, in order to clarify this phenomenon, more detailed studies on the prepolymerization would be necessary.

Based on this investigation, the kinetic measurements were carried out with the prepolymerized catalyst with which highest activity was reached in order to exclude the possible limitations described above.

### 6.8.3 Results kinetic measurements

Based on the previous studies, the final kinetic measurements were carried out according procedure B with a prepolymerized catalyst with a DP of 160 $mg_{PP}/mg_{Cat}$. Aim of the measurements was to determine the influence of reaction temperature and hydrogen concentration on activity and to measure the kinetic profiles of the metallocene catalyst for the bulk polymerization of propylene. In the following, the derived avg. activities as well as the measures activity profiles are shown.

#### 6.8.3.1 Influence of reaction temperature

Polymerizations were carried out in the temperature range from 55°C to 80°C. Figure 75 shows the average activities (a) as well as the corresponding activity profiles (b) for each reaction temperature exemplarily for a hydrogen feed concentration of 0.11 mol%.

a)

b)

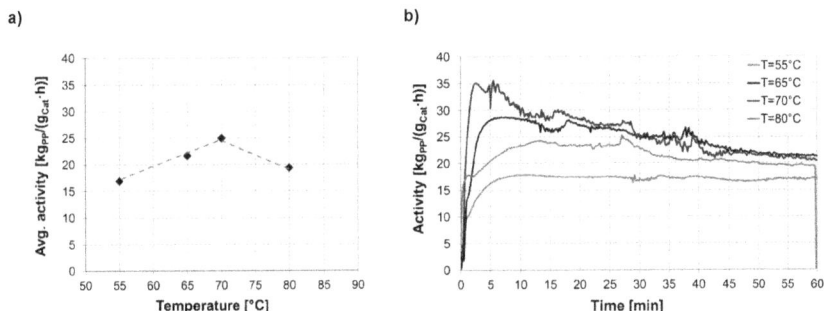

Figure 75: Results bulk polymerization: Influence of reaction temperature on a) average activities, b) activity profiles (T=55-80°C, $c_{H2,feed}$=0.11 mol%)

With increasing reaction temperatures from 55°C to 70°C the average activity increases. The maximum activity is reached at 70°C. A further increase above 70°C leads to a decrease in activity.

The corresponding activity profiles are showing the same trend: The activity level increases with increasing reaction temperature from 55°C to 70°C. After the injection of the prepolymerized catalyst (t=0 min), the activity increases fast whereby the maximum initial activity increases and is reached faster with increasing reaction temperature. During reaction the activity decreases, whereby the deactivation is stronger with increasing reaction temperature. For a higher reaction temperature of 80°C, the reached activity level is lower compared to those at 65°C and 70°C and also showing a slightly different course. After reaction start, the activity increases slower reaching a maximum activity after approx. 13 min. During polymerization, a decrease of the activity over reaction time is also visible.

The same trends for the avg. activities as well as for the activity profiles were derived for polymerizations with the other $H_2$ concentrations. Figure 76 shows the reached avg. activities for the different $H_2$ concentration over the investigated temperature range. From 55°C to 70°C the average activity generally increases reaching a maximum at 70°C. A further increase above 70°C leads to a decrease in activity. An exception is the polymerization with 0.03 mol% $H_2$ where a possible maximum activity could be expected between 70°C and 80°C. The corresponding activity profiles at the different reaction temperatures are shown in Figure 79, page 118.

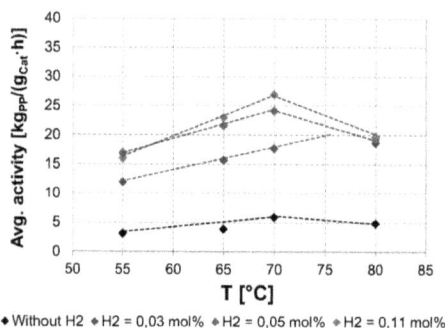

Figure 76: Results bulk polymerization: Influence of reaction temperature on avg. activity for all investigated $H_2$ concentrations (T=55-80°C, $c_{H2,feed}$=0-0.11 mol%)

## 6.8.3.2 Influence of hydrogen

In order to study the effect of $H_2$ on the catalyst activity, polymerizations were carried out with $H_2$ concentrations ranging from 0 to 0.11 mol% at the different reaction temperatures from 55 to 80°C and, in addition, with 0.14 mol% $H_2$ for a reaction temperature of 70°C (note: all $H_2$ concentrations corresponds to initial $H_2$ feed concentrations). Figure 77 shows the average activities (a) as well as the corresponding activity profiles (b) for each $H_2$ concentration exemplarily for a reaction temperature of 70°C.

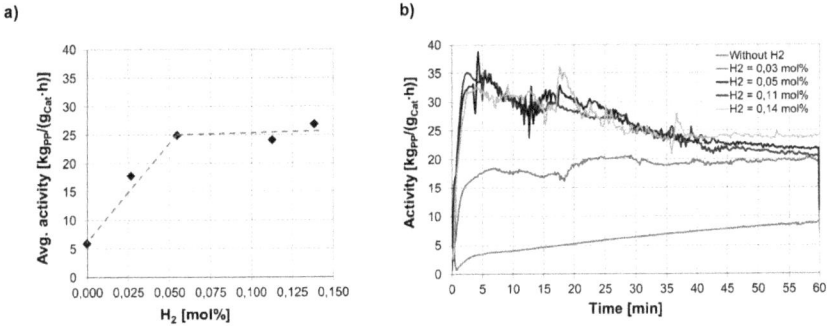

Figure 77: Results bulk polymerization: Influence of hydrogen on a) average activity b) activity profiles
(T=70°C, $c_{H2,feed}$=0-0.14 mol%)

With increasing $H_2$ concentration from 0 to approx. 0.05 mol% the average activity strongly increases. A further increase above 0.05 mol% $H_2$ leads to no further increase of the avg. activities. For $H_2$ concentrations between 0.05 to 0.14 mol% an activity plateau with similar avg. activities is reached.

The same trend is also visible in the activity profiles. At higher $H_2$ concentrations above 0.05 mol% the measured activity profiles are quite similar and showing the same course over reaction time. After the reaction start, the activity increases fast reaching a maximum activity after approx. 3 to 5 min. With further reaction progress, a similar deactivation behavior is visible. With 0.025 mol% $H_2$ the activity level is lower and no clear maximum activity and deactivation profile is visible. Without $H_2$ the reaction start is significant slower compared to reactions with $H_2$. After 60 min, the activity is still increasing. A deactivation might be occur but is not visible in the measured time.

The influence of the different $H_2$ concentrations on the activity was also measured at the different reaction temperatures. The resulting avg. activities are shown in Figure 78. Also here, a similar effect of hydrogen on the avg. activity is observed at the different reaction temperatures. For reaction temperatures from 55°C to 70°C the activity plateau is reached with approx. 0.05 mol% $H_2$. At higher reaction temperature of 80°C the activity plateau is already reached with 0.025 mol% $H_2$. An explanation could be that at 80°C maybe more $H_2$ was solved in the liquid phase. Calculations with Aspen showed that with an initial $H_2$ feed of 0.025 mol% at 80°C the $H_2$ concentration in liquid phase ($H_{2,lq.}$) is with 0.019 mol% higher compared to $H_2$ lq. concentration at lower temperatures (e.g. 70°C, $H_{2,lq}$=0.015 mol%).

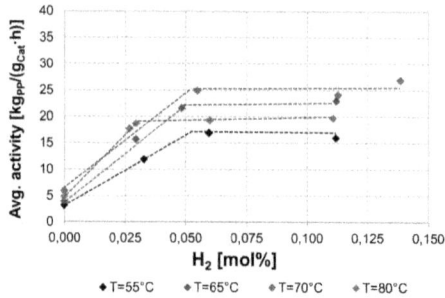

Figure 78: Results bulk polymerization: Influence of hydrogen on average activity for all reaction temperatures (T=55-80°C, $cH_{2,feed}$=0-0.14 mol%)

The corresponding activity profiles at the different reaction temperatures are shown in Figure 79. It can be seen that the course of the activity profiles for polymerizations at 55°C and 65°C are similar to the activity profile at 70°C but at a lower activity level. For a reaction temperature of 80°C the courses of activities show a slower activation without a clear activity maximum. Similar activity profiles are derived for $H_2$ concentrations already at 0.03 mol% (rounded up from 0.025 mol%). For polymerizations without $H_2$ the activity slowly increases during the course of the polymerization and no clear activity maximum as well as deactivation is visible for all reaction temperatures.

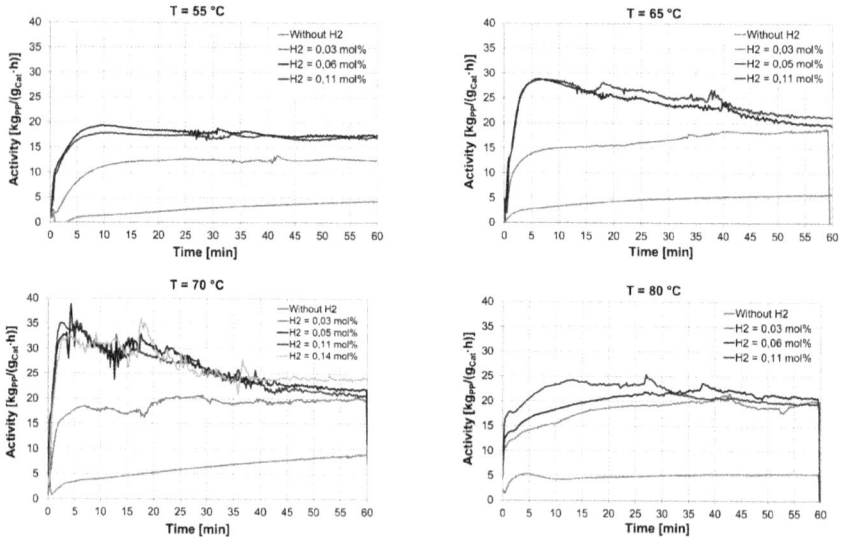

Figure 79: Results bulk polymerization: Influence of hydrogen on activity profiles for all investigated reaction temperatures (T=55-80°C, $cH_{2,feed}$= 0-0.14mol%)

118

### 6.8.4 Results analytical measurements

In the following, the influences of the different reaction conditions (reaction temperature and $H_2$ concentration) on the polymer characteristics are discussed. A summary of the detailed results is given in Table 35 and Table 36 in appendix 9.2.1.

#### 6.8.4.1 Molecular weights and molecular weight distribution

Molecular weights and molecular weight distribution of the produced polymer samples were analyzed with high temperature GPC by the external cooperation partner.

The influence of hydrogen on the molecular weight is shown exemplary for a polymerization at 80°C in Figure 80. As expected, with increasing $H_2$ concentration $M_w$ decrease. The same trend was also observed for the other reaction temperatures (see Table 35, appendix 9.2.1). Figure 80 also shows the calculated PDIs. The PDI for polymerizations without $H_2$ could not be determined accurately from the measured GPC curves and are therefore not representative. For polymer samples produced with different $H_2$ concentrations, similar PDIs were derived. $H_2$ has therefore no significant influence on the broadness of the MWD. The calculated average PDI for polymers produced at 80°C is 3.32. The average PDIs for polymers produced at the other reaction temperatures are with values from 3.01 to 3.45 in the same range. The derived PDI are slightly higher as normally expected - polymers produced with MC catalyst have a very narrow MWD with a PDI of approx. 2 [14]. A possible explanation for the slightly higher PDI might be that the use of a supported MC catalyst leads to a slight broadening of the MWD [50].

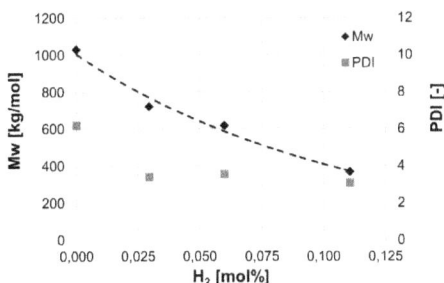

Figure 80: Results bulk polymerization – Analytics: Influence of hydrogen on weight average molecular weight and polydispersity index (T=80°C, $cH_{2,feed}$= 0-0.11mol%)

#### 6.8.4.2 Correlation $M_w$ and MFR

The melt mass flow rates of the produced polymer samples were measured in our laboratory using a micro melt flow indexer (see chapter 4.5.1). The MFR generally gives a fast indication of the melt viscosity and, thus, also the range of the molecular weight of the polymer. Figure 81 shows a correlation between the MFR values and weight average molecular weights determined by GPC measurements. The shown correlation could be used to determine $M_w$ of polymer samples where no GPC measurements were performed.

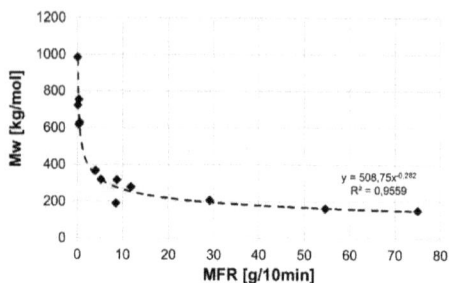

Figure 81: Results bulk polymerization – Analytics: Correlation weight average molecular weight and MFR value

## 6.8.4.3 Results crystallinity, porosity and density measurements

The influence of different reaction conditions on the polymer characteristics was generally studied for polymer samples produced at temperatures ranging from 55 to 80°C with 0.05 mol% $H_2$ and at 70°C with 0 to 0.14 mol% $H_2$ ($H_2$ concentrations correspond to initial $H_2$ feed). A detailed overview of the results is given in Table 36 in appendix 9.2.1.

Crystallinity

The crystallinity of the polymer samples was determined by DSC measurements (chapter 4.5.2). Figure 82 shows the resulting crystallinities of polymer samples (analyzed from the first melting peak) produced at different temperature and hydrogen concentrations. It can be seen that similar crystallinities were obtained at different reaction temperatures, whereas a slight increase of crystallinity can be recognized with increasing $H_2$ concentration. In general, the polymer produced with the supported metallocene catalyst has an average crystallinity of 46.6 %.

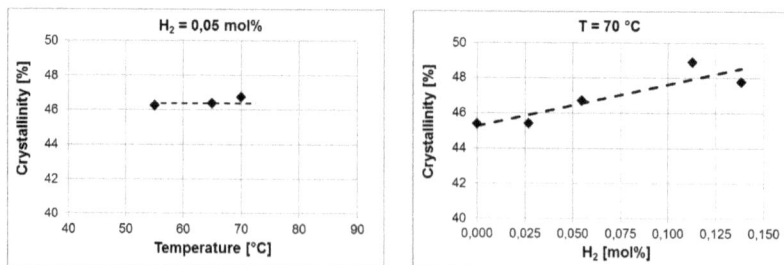

Figure 82: Results bulk polymerization – Analytics: Crystallinity left) influence of reaction temperature, right) influence of hydrogen concentration

## Porosity

The porosity was measured with mercury porosimetry as described in chapter 4.5.3. The influence of reaction temperature and $H_2$ concentration on the particle porosity is shown in Figure 83. It can be seen that the porosity of the polymer particle decreases with increasing reaction temperature. Polymers produced at same reaction temperature but with different $H_2$ concentrations showing no significantly different porosity. It has to be mentioned that the results show good tendencies of porosity in respect to temperature and $H_2$ concentrations, they could also be reproduced. But care should be taken using absolute values due to uncertainties during the porosity measurements (influence of applied pressure on pore diameters).

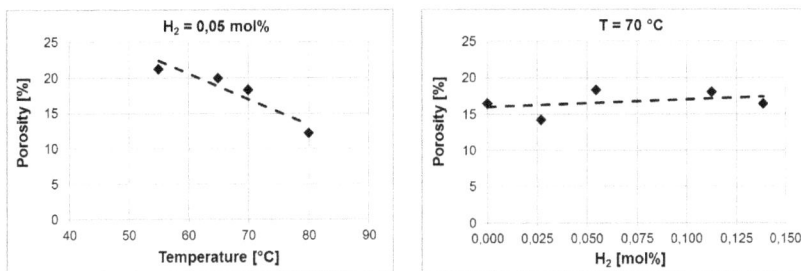

Figure 83: Results bulk polymerization – Analytics: Porosity left) influence of reaction temperature, right) influence of hydrogen concentration

## Bulk density

The bulk density was determined as described in chapter 4.5.4. The results are shown in Figure 84. For reaction temperatures ranging from 55°C to 70°C the produced polymer powders have similar bulk densities of approx. 0.43 g/ml. A slightly lower bulk density of 0.40 g/ml was obtained for polymer powder produced at 80°C. This could be a hint that the size or shape of the polymer particles was influenced by the higher reaction temperature. The bulk densities of polymer samples produced with different $H_2$ show also no significant differences.

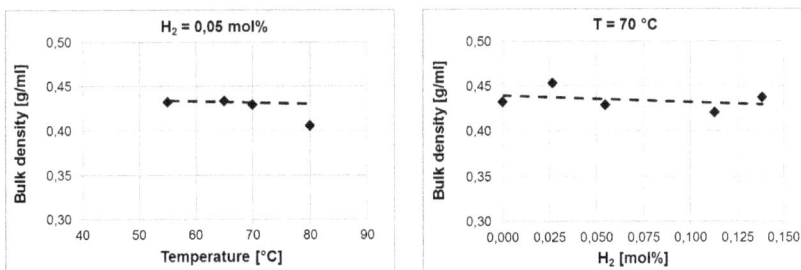

Figure 84: Results bulk polymerization – Analytics: Bulk density left) influence of reaction temperature, right) influence of hydrogen concentration

121

## 6.8.4.4 Electron microscopy

In order to investigate the influence of the reaction conditions on the polymer morphology, pictures of the polymer samples were analyzed using scanning electron microscopy (SEM, chapter 4.5.6).

Figure 85 shows a comparison of the catalyst, the prepolymer and the resulting polymer powder. The supported metallocene catalyst particles have a relatively spherical shape. During prepolymerization, the polymer grows and a prepolymer also with a spherical shape is formed. No particle breakage is visible which can be related to the mild catalyst injection conditions. When the prepolymerized catalyst was then injected at reaction temperatures, the particle morphology could be preserved; spherical polymer particles are obtained showing smooth surfaces and no particle break-up.

SEM pictures for different polymerization temperatures (Figure 86) and different hydrogen concentrations (Figure 87) do show qualitatively the same behavior. In the studied range, these parameters do not influence particle morphology, apart from particle size due to different productivities.

| Magnification | Catalyst | Prepolymer | Polymer |
|---|---|---|---|
| 15x (1mm) | | | |
| 40x (500µm) | | | |

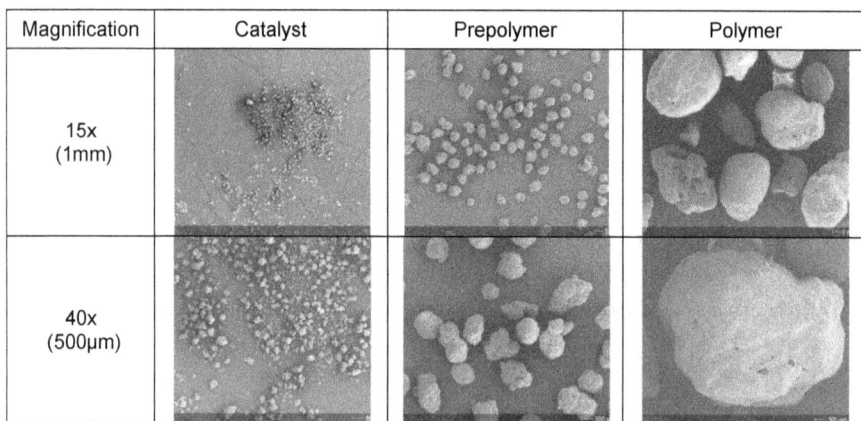

Figure 85: Results bulk polymerization – Analytics: Comparison particle morphology of supported metallocene catalyst, prepolymer and polymer (V87, T=70°C, $c_{H_2}$=0.05mol%)

| Magnification | T = 55°C (V86) | T = 70°C (V87) | T = 80°C (V100) |
|---|---|---|---|
| 15x (1mm) | | | |
| 40x (500µm) | | | |

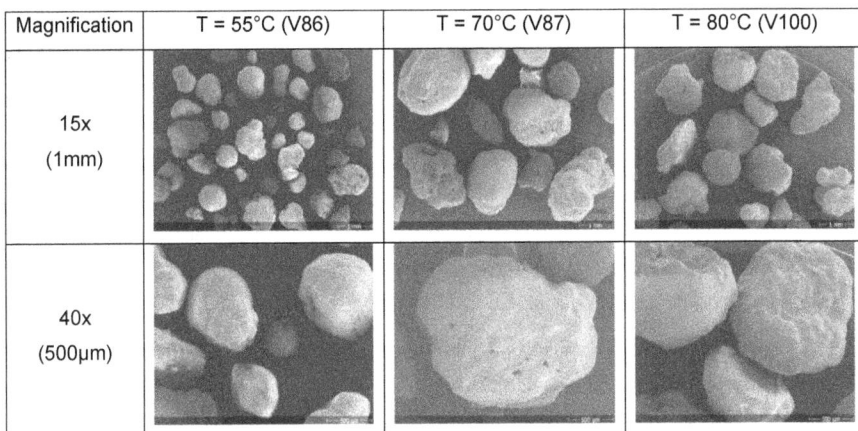

Figure 86: Results bulk polymerization – Analytics: Influence of reaction temperature on polymer morphology ($cH_2$= 0.05 mol%)

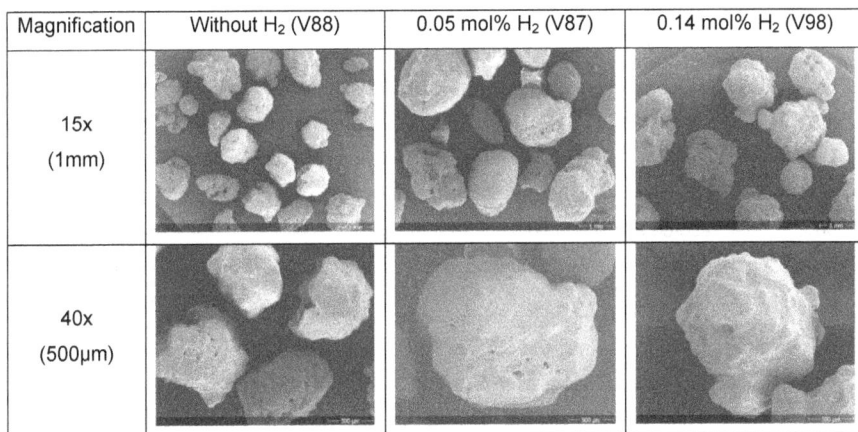

| Magnification | Without $H_2$ (V88) | 0.05 mol% $H_2$ (V87) | 0.14 mol% $H_2$ (V98) |
|---|---|---|---|
| 15x (1mm) | | | |
| 40x (500µm) | | | |

Figure 87: Results bulk polymerization – Analytics: Influence of hydrogen on polymer morphology (T=70°C)

## 6.9 Summary experimental investigation of the bulk polymerization of propylene with a supported metallocene catalyst

Kinetic measurements of a supported metallocene catalyst were carried out in liquid propylene using a special heat flow reaction calorimeter. Aim of the study was to investigate bulk polymerization kinetics at different reaction conditions (different reaction temperatures and $H_2$ concentrations). A particular focus was to study the influence of prepolymerization on the resulting catalyst activity. The produced polymer powder was analyzed with different analytical methods.

The experimental setup was build up during the work. An improved raw material purification system was developed which is crucial for obtaining good activities with sensible metallocene catalysts. Two different polymerization procedures (polymerization with in-situ prepolymerization and with external prepolymerization) were developed. With both procedures same activities as well as MFR values could be obtained with high reproducibility. The procedure with external prepolymerization has the advantage of more defined prepolymerization conditions since there is no undefined heating-up period and the degree of prepolymerization is exactly known. Furthermore, with this method the kinetic data are earlier accessible. Hence, the procedure with external prepolymerization was chosen for further studies.

The investigation of the influence of different prepolymerization degrees on the resulting avg. activity showed that maximum activity was reached using a prepolymer with a DP of 160 $mg_{PP}/mg_{Cat}$ and was therefore used for final kinetic measurements.

The kinetic measurements were carried out in the temperature range from 55°C to 80°C and with hydrogen feed concentrations from 0 to 0.14 mol%. Following main conclusions are derived from the investigations:

At constant $H_2$ concentration, avg. activity increased with increasing reaction temperatures from 55°C to 70°C. A further increase of the reaction temperate above 70°C led to a decrease in activity. The same trend could be observed for all $H_2$ concentrations, except for polymerizations with approx. 0.03 mol% $H_2$, where a possibly activity maximum is expected between 70°C and 80°C. The measured activity profiles followed the same trends: the activity levels increased with increasing temperature up to 70°C and decreased at higher reaction temperature of 80°C (exception for 0.03 mol% $H_2$).

At constant reaction temperature of 55°C to 70°C, an increase of the avg. activity was observed with increasing $H_2$ concentrations from 0 to 0.05 mol% where an activity plateau was reached. A further increase above 0.05 mol% led to same avg. activities. For higher reaction temperature of 80°C the activity plateau was reached already with approx. 0.03 mol% $H_2$ due to a higher concentrations of $H_2$ in liquid phase. The level of the activity profiles followed the same trends.

When $H_2$ was present, the general course of the activity showed a fast increase of the activity at the beginning of the reaction where a maximum was reached. During polymerization, the activity decreased. The maximum activity reached at the beginning of the reaction and the deactivation behavior over time was stronger with increasing reaction temperature from 55°C to 70°C and with increasing $H_2$ (until $H_2$ concentration where the activity plateau was reached). At 80°C, no clear maximum during the course of the polymerization could be seen. Also a clear deactivation was not visible or started with a delay. In contrast, without $H_2$ the reaction started slow without showing an activity maximum as well as a typical deactivation behavior within the reaction time. Similar kinetic profiles were derived at the different reaction temperatures.

# 7 Kinetic modeling of the bulk polymerization with a supported metallocene catalyst

Focus of this chapter is the derivation of a phenomenological model describing the polymerization kinetics of the bulk polymerization of propylene with a supported (prepolymerized) metallocene catalyst. Based on the experimental derived data (avg. activities, activity profiles and molecular weights), a simplified kinetic scheme is developed and the kinetic parameters of the metallocene catalyst are estimated. Simulations and parameter estimation were carried out using the software gProms ModelBuilder (Process Systems Enterprise).

In chapter 7.1, the model assumptions and the kinetic scheme of the bulk polymerization as well as the calculations for monomer and hydrogen concentrations in the liquid phase are described. The results of the parameter estimation, the comparison between experimental and simulated activity profiles and avg. molecular weights as well as the estimated kinetic parameters are presented in chapter 7.2.

## 7.1 Derivation of the kinetic model for the bulk polymerization

### 7.1.1 Model assumptions

Metallocene catalysts belong to the single site catalysts (see chapter 2.1.2) and therefore only one sort of active sites is considered within this model. The kinetic scheme is developed on the basis of measured activity profiles from polymerizations carried out with prepolymerized catalyst. Due to the larger heat transfer area of the prepolymerized catalyst as well as the better heat transfer conditions from the polymer particle to the surrounding liquid bulk phase in comparison to the gas-phase, neither heat transfer limitations nor particle overheating are considered. The mass balances are derived for a quasi-homogeneous particle model, wherein mass transfer limitations as well as swelling of the polymer are neglected. Average molecular weights are calculated using the method of moments. The temperature dependency of rate constants is described by the Arrhenius equation. The equilibrium monomer concentration at the active sites of the catalyst is calculated according the Flory-Huggins equation and is described in chapter 7.1.4. The hydrogen concentration within the liquid phase is calculated by means of Henry's law and is discussed in chapter 7.1.5.

### 7.1.2 Kinetic scheme

The simplified kinetic scheme for the polymerization reaction with metallocene catalyst is shown in Figure 88. In the initiation reaction (reaction I), the already activated catalyst (denoted as Zr*) reacts with the monomer forming polymer with chain length one. Chain growth is described in the propagation reaction (reaction II), wherein the monomer is inserted into the polymer chain with chain length n resulting in polymer with chain length n+1. As described in chapter 2.3, hydrogen has two functions in the polymerization reaction: Firstly, it acts as chain transfer agent influencing the molecular weight of the polymer and secondly, in case of propylene polymerization, it accelerates the polymerization reaction. In the experimental investigation it could be shown that an increase of the $H_2$ concentration leads to an increase of the activity until a certain $H_2$ concentration level. According to literature, this phenomenon can be described with the dormant site theory (chapter 2.3 and 2.2.2). In the kinetic model, formation of dormant sites and its reactivation with $H_2$ is described with reaction (III) and (IV). In case of polymerizations without $H_2$, an additional reaction step for the reactivation of dormant sites to active sites is necessary (reaction V). In order to describe the molecular weights two transfer reactions are considered: The chain transfer to hydrogen (reaction VI) and spontaneous chain transfer by ß-hydride elimination (reaction VII). The deactivation of the catalyst is described by a spontaneous deactivation. The following reaction scheme results:

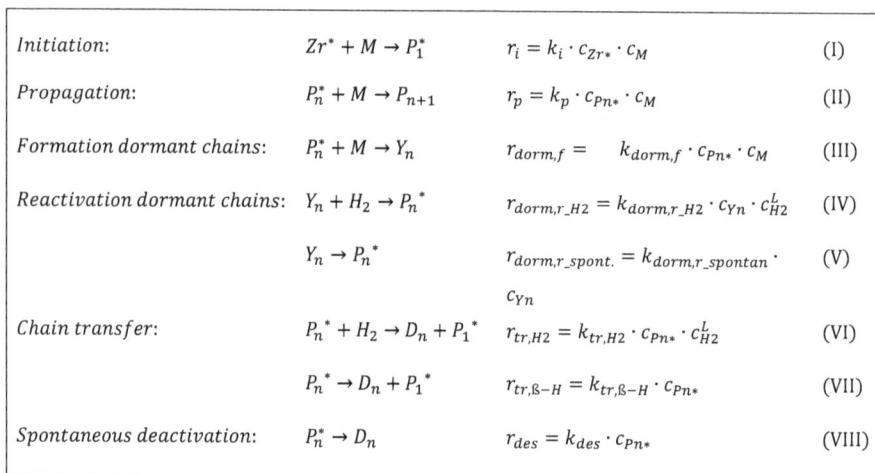

| | | | |
|---|---|---|---|
| *Initiation:* | $Zr^* + M \rightarrow P_1^*$ | $r_i = k_i \cdot c_{Zr*} \cdot c_M$ | (I) |
| *Propagation:* | $P_n^* + M \rightarrow P_{n+1}$ | $r_p = k_p \cdot c_{Pn*} \cdot c_M$ | (II) |
| *Formation dormant chains:* | $P_n^* + M \rightarrow Y_n$ | $r_{dorm,f} = k_{dorm,f} \cdot c_{Pn*} \cdot c_M$ | (III) |
| *Reactivation dormant chains:* | $Y_n + H_2 \rightarrow P_n^*$ | $r_{dorm,r\_H2} = k_{dorm,r\_H2} \cdot c_{Yn} \cdot c_{H2}^L$ | (IV) |
| | $Y_n \rightarrow P_n^*$ | $r_{dorm,r\_spont.} = k_{dorm,r\_spontan} \cdot c_{Yn}$ | (V) |
| *Chain transfer:* | $P_n^* + H_2 \rightarrow D_n + P_1^*$ | $r_{tr,H2} = k_{tr,H2} \cdot c_{Pn*} \cdot c_{H2}^L$ | (VI) |
| | $P_n^* \rightarrow D_n + P_1^*$ | $r_{tr,\text{ß}-H} = k_{tr,\text{ß}-H} \cdot c_{Pn*}$ | (VII) |
| *Spontaneous deactivation:* | $P_n^* \rightarrow D_n$ | $r_{des} = k_{des} \cdot c_{Pn*}$ | (VIII) |

Figure 88: Kinetic scheme for the bulk polymerization with metallocene catalyst

## 7.1.3  Derivation of mass balances

In analogy to chapter 5.1.5, the mass balances for the individual components are derived from the kinetic scheme. As described for the kinetic modeling of the gas-phase polymerization in chapter 5.1.2, the increase of the reaction volume $V_R$ (polymer phase) during polymerization would lead to a decrease of the concentration of active sites of the growing chains. This "dilution effect" is therefore considered by balancing the number of moles of the component as a product of reaction volume and its concentration. According the developed kinetic scheme following mass balance for the growing polymer chains $P_n^*$ is obtained:

$$\frac{dn_{Pn*}}{dt} = V_R\left(r_i + r_{dorm,r\_H2} + r_{dorm,r\_spontan} - r_{dorm,f} - r_{des}\right) \tag{7.1}$$

$$= k_i \cdot c_{C3}^* \cdot n_{Zr*} + n_{Yn}\left(k_{dorm,r\_H2} \cdot c_{H2}^L + k_{dorm,r\_spontan}\right) - n_{Pn*}\left(k_{dorm,f} \cdot c_{C3}^* + k_{des}\right)$$

The mass balances for the other components, the active component $Zr^*$, the dormant chains $Y_n$ and the dead polymer chains $D_n$ are derived in the same way:

$$\frac{dn_{Zr*}}{dt} = -k_i \cdot n_{Zr*} \cdot c_{C3}^* \tag{7.2}$$

$$\frac{dn_{Yn}}{dt} = k_{dorm,f} \cdot n_{Pn*} \cdot c_{C3}^* - k_{dorm,r\_H2} \cdot n_{Yn} \cdot c_{H2}^L - k_{dorm,r\_spontan} \cdot n_{Yn} \tag{7.3}$$

$$\frac{dn_{Dn}}{dt} = k_{tr,H2} \cdot n_{Pn*} \cdot c_{H2}^L + k_{tr,\text{ß}-H} \cdot n_{Pn*} + k_{des} \cdot n_{Pn*} \tag{7.4}$$

The initial number of moles of active component $n_{Zr*}^0$ is calculated from the mass of catalyst and the weight fraction of the active component divided by its molecular weight:

$$n_{Zr*}^0 = \frac{m_{Cat} \cdot w_{Zr}}{MW_{Zr}} \tag{7.5}$$

The initial number of moles of growing chains, dormant chains and dead chains is equal to zero.

$$n_{Pn*}^0 = n_{Yn}^0 = n_{Dn}^0 = 0 \tag{7.6}$$

The activity is finally calculated according following equation:

$$A = \frac{k_p \cdot n_{Pn*} \cdot c_{C3}^* \cdot MW_{C3}}{m_{Cat}} \cdot \frac{3600\frac{s}{h}}{1000\frac{g}{kg}} \qquad \left[\frac{kg_{PP}}{g_{cat} \cdot h}\right] \tag{7.7}$$

The temperature dependency of the rate constants of the propagation reaction $k_p$, transfer reactions $k_{tr,H2}$ and $k_{tr,\text{ß}-H}$ as well as the deactivation reaction $k_{des}$ is described by the Arrhenius equation (see equations (5.4) to (5.7), chapter 5.1.2). The rate constants for the formation and reactivation of dormant chains ($k_{dorm,f}$, $k_{dorm,r\_H2}$ and $k_{dorm,r\_spontan}$) are assumed to be temperature independent as no clear distinction of temperature dependency of propagation and the formation and reactivation of dormant sites could be determined from the experimentally derived kinetic profiles.

Average molecular weights are calculated according to the method of moments. The differential equations for the zeroth, first and second moment of the three polymer species are derived from the developed kinetic scheme in the same manner as described in chapter 5.1.7 and are shown in Appendix 9.2.2. Weight average molecular weight $M_w$, number average molecular weight $M_n$ and PDI are calculated according the equations (5.49) to (5.51) given in chapter 5.1.7.

### 7.1.4 Determination of the monomer concentration in the polymer particle

For determination of the equilibrium monomer concentration in the polymer particle $c_{C3}^*$ used in the kinetic expressions (7.1) to (7.7) it is assumed that:

- overall monomer conversion is that low that liquid propylene is still present, which is a precondition for bulk polymerization
- the concentration of monomer in the polymer is in equilibrium to the liquid propylene phase surrounding the polymer particles.

The equilibrium solubility of propylene in polypropylene is calculated via the Flory-Huggins equation, which describes the solubility of low molecular components in an (amorphous) polymer [145].

$$ln\left(\frac{p}{p^0}\right) = ln\phi + (1 - \phi) + \chi(1 - \phi)^2$$

$$(7.8)$$

Therein, $p$ and $p^0$ are the partial pressure and the saturation vapor pressure of the monomer, respectively, $\phi$ is the volume fraction of the permeant (monomer) sorbed into the amorphous parts of the polymer and $\chi$ is the Flory-Huggins interaction parameter.

$\phi$ is defined as:

$$\phi = \frac{V_M}{V_M + V_{PP}}$$

$$(7.9)$$

where $V_M$ is the volume fraction of the monomer and $V_{PP}$ is the amorphous polymer in the polymer/monomer mixture. As the polymerization occurs only in the amorphous parts of the polymer particle, the equilibrium monomer concentration $c_{C3}^*$ required in the kinetic expressions (denoted as $c_M$ in the kinetic scheme Figure 88) can be calculated according following equation [67].

$$c_{C3}^* = \phi \cdot c_{C3}^L$$

$$(7.10)$$

with $c_{C3}^L$ as the concentration of the liquid monomer surrounding the polymer particle.

The interaction parameter $\chi$ can either be determined experimentally by fitting sorption measurements [73],[80],[91] or calculated with the van Laar-Hildebrand equation [92],[67]. As $\chi$ does not only depend on reaction temperature but also on crystallinity of the polymer, absolute values from literature cannot simply be adopted. In this work $\chi$ was therefore estimated using the van Laar-Hildebrand equation [146]:

$$\chi = \left(\frac{v_M}{R \cdot T}\right) \cdot (\delta_M - \delta_{PP})^2 + \beta$$

$$(7.11)$$

where $v_M$ is the molar volume of the liquid monomer, $\delta_M$ and $\delta_{PP}$ are the solubility parameters of the monomer and polymer, respectively, and $\beta$ is the lattice constant. It has to be mentioned that the universal gas constant $R$ must be used in $cal/(mol \cdot K)$.

The molar volume is calculated from the molecular weight $MW_{C3}$ and the density of liquid monomer $\rho_{C3}^L$ at the given temperature:

$$v_m = \frac{MW_{C3}}{\rho_{C3}^L} \tag{7.12}$$

The solubility parameters $\delta_M$ and $\delta_{PP}$ are calculated using the correlations given by Bradford and Thodos [147] and Hutchinson and Ray [99]:

$$\delta_M = \delta_c + k \cdot (1 - T_R)^m \tag{7.13}$$

$$\delta_{PP} = 7.7 - 0.0121 \cdot (T - 303.15) \tag{7.14}$$

with $\delta_c$ as the solubility parameter at the critical point, $T_R$ as reduced temperature which is defined as $(T/T_{crit})$, and $k$ and $m$ as constants. The values of these constants for polypropylene are summarized in Table 23 together with the other constants needed for the calculation of $\chi$.

Table 23: Data for the calculation of the interaction parameter $\chi$ according the van Laar-Hildebrand equation

| Constant | Value | Unit | Reference |
|---|---|---|---|
| $\delta_c$ | 2.482 | $(cal/cm^3)^{0.5}$ | Bradford and Thodos [147] |
| $k$ | 7.65 | $(cal/cm^3)^{0.5}$ | Bradford and Thodos [147] |
| $m$ | 0.447 | - | Bradford and Thodos [147] |
| $\beta$ | 0.34 | - | Aminabhavi [148] |
| $T_{crit}$ | 365.57 | K | NIST database [132] |
| $R$ | 1.98589 | cal/(mol·K) | |

Results calculation of the monomer concentration

At saturation pressure, which corresponds to bulk conditions, $p/p^0$ in the Flory-Huggins equation (7.8) becomes one. As described above, $\chi$ was calculated using the van Laar - Hildebrand equation. The volume fraction $\phi$ in the Flory-Huggins equation could be then determined iterative using the Newton method. Table 24 shows the main results for the calculation of the monomer concentration in the polymer particle.

Table 24: Results for the calculation of the monomer concentration $c_{C3}^*$ in the polymer particle at bulk conditions ($p/p^0 = 1$)

| $T$ | $\rho_{C3}^L$ | $v_M$ | $\delta_M$ | $\delta_{PP}$ | $\chi$ | $\phi$ | $c_{C3}^L$ | $c_{C3}^*$ |
|------|------|------|------|------|------|------|------|------|
| °C | g/l | l/mol | (cal/cm³)^0.5 | (cal/cm³)^0.5 | - | - | mol/l | mol/l |
| 55 | 444.39 | 0.095 | 5.24 | 7.40 | 1.01 | 0.31 | 10.56 | 3.25 |
| 65 | 418.85 | 0.100 | 4.89 | 7.28 | 1.20 | 0.22 | 9.95 | 2.23 |
| 70 | 404.12 | 0.104 | 4.68 | 7.22 | 1.32 | 0.18 | 9.60 | 1.74 |
| 80 | 367.50 | 0.115 | 4.17 | 7.10 | 1.74 | 0.099 | 8.73 | 0.86 |

With increasing reaction temperature the monomer concentration in the polymer particle decreases. Furthermore, the monomer concentration in the polymer particle is much lower compared to the bulk phase surrounding the polymer particle. The same correlations were already shown by Samson et al. [67] and Patzlaff [92], who used the van Laar-Hildebrand equation for their calculations. Also Hutchinson and Ray [99] concluded from their studies that monomer concentration in the polymer phase is lower as in the bulk phase, since the monomer exists already in the condensate state and no phase change occurs during the sorption process. They also stated that the ratio of monomer concentration in the polymer phase to the monomer concentration in the bulk phase decreases with increasing reaction temperature. Nevertheless, Meier et al. [73] reported a different behavior. They found an increase of the monomer concentration in the polymer particle with increasing reaction temperature and calculated much higher monomer concentrations. Their calculations were based on sorption measurements with an experimental determination of the interaction parameter $\chi$. A comparison of the calculated monomer concentration with literature data is shown in Figure 89.

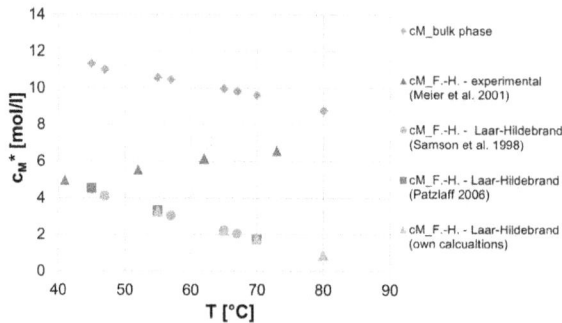

Figure 89: Comparison monomer concentration in polymer particle with literature data [67],[73], [92]

## 7.1.5 Determination of hydrogen concentration in liquid phase

As shown in the experimental studies, hydrogen influences the molecular weight of the polymer as well as the kinetics of the polymerization reaction. Therefore, hydrogen concentration in liquid propylene needs to be considered according to the reaction conditions.

According to the kinetic scheme, hydrogen is consumed in the transfer reaction and in the reactivation of sites. Thus, the overall moles of hydrogen in the reactor change according to:

$$\frac{dn_{H2}^{total}}{dt} = -k_{tr,H2} \cdot n_{Pn*} \cdot c_{H2}^L - k_{dorm,r\_H2} \cdot n_{Yn} \cdot c_{H2}^L \tag{7.15}$$

However, the partitioning of hydrogen between liquid phase and gas-phase has to be considered, since the filling level can change significantly during a bulk-phase polymerization (Figure 90):

$$n_{H2}^{total} = c_{H2}^L \cdot V_L + \frac{p_{H2} \cdot V_G}{RT} \tag{7.16}$$

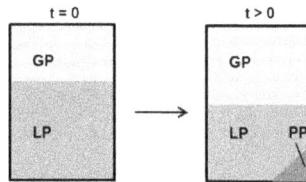

Figure 90: Schematic illustration of volume changes during bulk polymerization

For hydrogen partitioning, thermodynamic equilibrium between gas-phase and liquid-phase, is considered. In open literature only limited data on the vapor-liquid equilibrium (VLE) of propylene and hydrogen is available. In 1954, Williams and Katz [149] presented VLE data of binary systems of hydrogen with several hydrocarbons such as propylene and ethylene. Mizan et al. [150] studied the solubility and mass transfer coefficients of hydrogen (and ethylene) in liquid propylene at a more relevant range of conditions and found that solubility of hydrogen can be described by Henry's law and that the solubility of $H_2$ increases with increasing temperature. Pater et al. [118] used gas chromatography to determine the VLE of hydrogen and liquid propylene and described their results using the Peng-Robinson equation of state. Meier et al. [73] used Henry's law and a correlation for the Henry constant based on the work of Mizan et al. [150] for calculations of the $H_2$ concentration in liquid and in gas-phase polymerization of propylene with a heterogeneous metallocene catalyst.

In this work, the hydrogen concentration in liquid propylene $c_{H2}^L$ was calculated using Henry's law:

$$c_{H2}^L = k_{H2}^* \cdot p_{H2}^G \tag{7.17}$$

with $k_{H2}^*$ as Henry coefficient in mol/(l·bar) and $p_{H2}^G$ as partial pressure of hydrogen.

The Henry coefficients $k_{H2}^*$ in equation (7.17) were derived by correlating the data given by Meier et al. [73]. Following correlation is used to calculate $k_{H2}^*$ for the investigated temperature range:

$$k_{H2}^* = 1{,}25 \cdot 10^{-6} \cdot \vartheta^2 + 2{,}935 \cdot 10^{-4} \cdot \vartheta + 6{,}745 \cdot 10^{-3} \tag{7.18}$$

with $\vartheta$ as reaction temperature in °C. The derived Henry coefficients are summarized together with the data used for calculation of the volume in Table 25. The densities of liquid and gaseous propylene were derived from NIST database [132]. For the density of polypropylene $\rho_{PP}$ an average polymer density of 886 g/l was used, which was measured by helium pycnometry (Pycnomatic ATC, Porotec GmbH) at the Martin-Luther-University Halle-Wittenberg.

Table 25: Relevant data for the calculation of the hydrogen concentration in liquid propylene

| T | $k_{H2}^*$ | $\rho_{C3}^L$ | $\rho_{C3}^G$ |
|---|---|---|---|
| °C | mol/(l·bar) | g/l | g/l |
| 55 | 0.0267 | 444.39 | 51.92 |
| 65 | 0.0311 | 418.85 | 67.13 |
| 70 | 0.0334 | 404.12 | 76.80 |
| 80 | 0.0382 | 367.5 | 103.36 |

Together with the closure-condition for volume (equation 7.19) and the overall mass-balance for unreacted and converted propylene (equation 7.20), the partitioning of propylene and hydrogen between gas and liquid phase and, thus, the hydrogen concentration in liquid phase can be calculated at any time.

$$V_R = V^G + V^L + \frac{m_{PP}}{\rho_{PP}} = const. \tag{7.19}$$

$$m_{C3}^{ges} = m_{C3}^G + m_{C3}^L + m_{PP} + m_{C3,PP} = \rho_{C3,l} \cdot V_L + \rho_{C3,g} \cdot V_G + m_{PP}\left(1 + \frac{c_{C3,PP} \cdot MW_{C3}}{\rho_{PP}}\right) = const. \tag{7.20}$$

## 7.2 Results kinetic modeling of the bulk polymerization with metallocene catalyst

In the following, the model implementation and parameter estimation are presented. The resulting kinetic parameters as well as the comparison of simulated with the experimentally derived activity profiles and weight average molecular weight are shown at the end of this chapter.

## 7.2.1 Model implementation and parameter estimation

For the simulations and the parameter estimation the software gProms ModelBuilder (Process Systems Enterprise Ltd.) was used. The model implementation was realized similar as described in the modeling of the gas-phase polymerization (chapter 5.1.9). The derived set of differential equations from the kinetic scheme, the moment equations of each species as well as the equations for the calculation of the hydrogen concentration and the activity were implemented in the "model" section. Data of the physical properties of propylene such as densities of liquid and gaseous phase and vapor pressures were taken from NIST database [132] (see Table 25). Catalyst relevant data such as the amount of active component on the catalyst particle were provided by the cooperation partner. Data of the polymer density and crystallinity were taken from the analytical measurements (see chapter 6.8.4.3). Also the calculated monomer concentration according to Flory-Huggins (see Table 24) was inserted for each reaction temperature.

In the "process" section, the kinetic constants (rate constants or pre-exponential factor and activation energy), which needed to be estimated, the reaction conditions (T, $m_{Cat}$, $m_{C3}$, $m_{H2}$) and the initial conditions of the differential equations were defined.

For the parameter estimation, the data of the measured activity-time profiles were prepared by using the current activity at selected times as described in the gas-phase polymerization. Furthermore, the experimentally derived weight average molecular weights of the polymer samples were included. The estimation of the kinetic parameters was performed with the gEst parameter estimation tool (see chapter 5.1.9) by fitting the model to the experimentally derived activity-time profiles and molecular weights, respectively. Due to the high number of the kinetic constants, the estimation was performed stepwise.

In a first step, rate constants of initiation, propagation and deactivation reaction were estimated by fitting the model with the experimental activity-time profiles for the different reaction temperatures at one hydrogen concentration (0.055 mol%). In case of the temperature dependent propagation and deactivation reaction, the rate constants were calculated by estimating the activation energy and the pre-exponential factor.

In a second step, the rate constants for the formation and reactivation of dormant sites were estimated. Herein, activity profiles of experiments with and without $H_2$ of all reaction temperatures were used.

The estimation of the parameters is an iterative process. With the first set of parameters for initiation, propagation and deactivation reaction, the second step was performed estimating the rate constants for formation and reactivation of dormant sites. With the new parameter set, step one was repeated followed by step two and so on. After each estimation step, the derived parameter set was revised carefully on plausibility.

In a third step, the rate constants of the transfer reactions (transfer to $H_2$ and ß-hydride elimination) were estimated by fitting the model to the experimental weight average molecular weights. Therein, as transfer reactions are considered to be temperature dependent, the activation energy and the pre-exponential factor were estimated.

## 7.2.2 Results parameter estimation

The final set of the estimated kinetic parameters of the supported metallocene catalyst is presented in Table 26.

Table 26: Kinetic modeling of the bulk polymerization with metallocene catalyst: Results parameter estimation

| Parameter | Estimated value | Unit |
|-----------|-----------------|------|
| $k_i$ | $2.87 \cdot 10^{-3}$ | $l/(mol \cdot s)$ |
| $k_p$ (70°C) | 8000 | $l/(mol \cdot s)$ |
| $E_{A,p}$ | 62.7 | $kJ/mol$ |
| $k_{tr\_H2}$ (70°C) | 778 | $l/(mol \cdot s)$ |
| $E_{A,tr\_H2}$ | 6.6 | $kJ/mol$ |
| $k_{tr\_ß-H}$ (70°C) | 0.65 | $1/s$ |
| $E_{A,tr\_ß-H}$ | 44.2 | $kJ/mol$ |
| $k_{des}$ (70°C) | $1.67 \cdot 10^{-4}$ | $l/(mol \cdot s)$ |
| $E_{A,des}$ | 71.6 | $kJ/mol$ |
| $k_{dorm,f}$ | 1.22 | $l/(mol \cdot s)$ |
| $k_{dorm,r\_H2}$ | 900 | $l/(mol \cdot s)$ |
| $k_{dorm,r\_spontan}$ | 0.2 | $1/s$ |

It has to be mentioned that different sets of parameters could describe the experimental data in comparable quality, which is an indication for potential cross-correlations between the estimated parameters. Nevertheless, the chosen parameter set gives the best description of the experimentally derived activity profiles and molecular weights at different reaction temperatures and different hydrogen concentrations explored so far. The estimated parameters for activation, propagation and deactivation constants for the propylene polymerization with metallocene catalyst are in a similar range to estimated parameters given in literature [71],[128].

## 7.2.3 Comparison of simulated and experimental data

### 7.2.3.1 Influence reaction temperature on activity

Figure 91 shows the activity profiles of the prepolymerized metallocene catalyst calculated with the model in comparison with the experimental data for the investigated temperature range from 55°C to 80°C exemplary for a hydrogen feed concentration of 0.11 mol%.

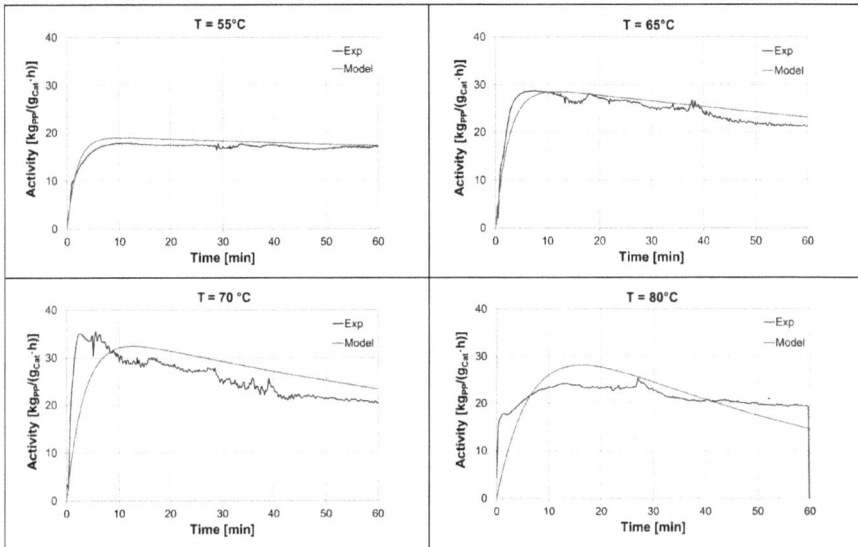

Figure 91: Results kinetic modeling bulk polymerization with metallocene catalyst: Comparison simulated and experimental data - Influence of reaction temperature on activity profiles (T=55-80°C, $H_{2,feed}$=0.11 mol%)

With the developed model, a reasonable fit between simulated and experimental activity profiles can be achieved. At higher reaction temperature of 70°C, the activity profile is slightly overestimated, whereas at 80°C the course of activity is slightly underestimated at the end of the reaction.

Comparing the simulated and experimental average activities for all reaction temperatures as well as for all investigated hydrogen concentrations in a parity diagram (Figure 92), it can be seen that the average activities can be calculated with the developed model within a 10% error range.

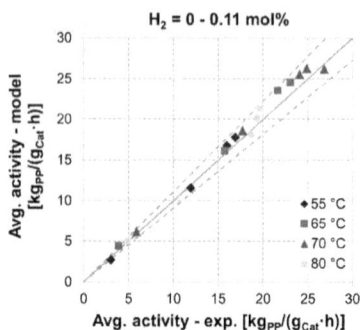

Figure 92: Results kinetic modeling bulk polymerization with metallocene catalyst: Parity diagram simulated vs. experimentally derived average activities (T=55-80°C, H$_{2,feed}$=0-0.11 mol%)

## 7.2.3.2 Influence H$_2$ concentration on activity

Figure 93 shows the simulated and experimentally derived activity profiles for the different hydrogen feed concentrations from 0 to 0.11 mol% exemplary for a reaction temperature of 70°C.

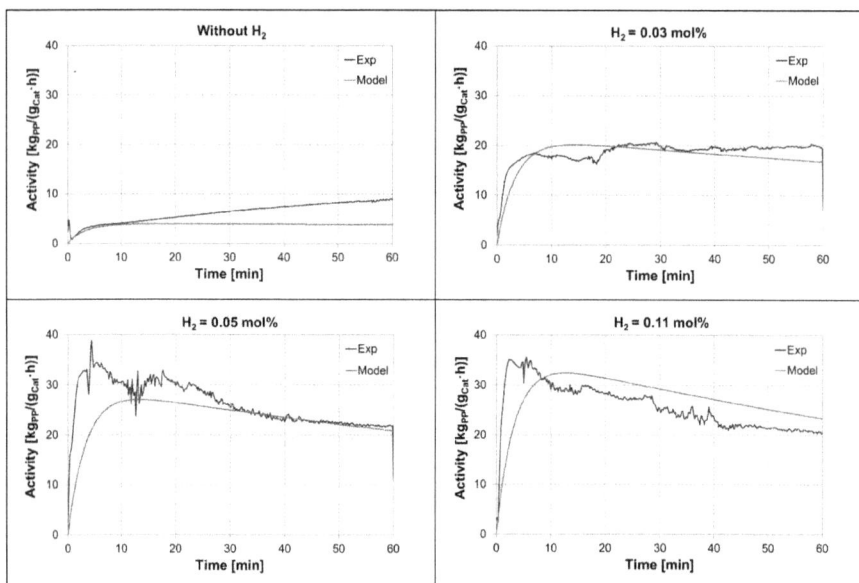

Figure 93: Results kinetic modeling bulk polymerization with metallocene catalyst: Comparison simulated and experimental data - Influence of hydrogen concentration on activity profiles (T=70°C, H$_{2,feed}$=0-0.11 mol%)

In case of the polymerizations with hydrogen, the model can predict the activity profile in an adequate manner during the course of the polymerization. For polymerization without hydrogen, a clear increase of the experimentally derived activity during the reaction can be seen, which could not be simulated with the model. One hypothesis for explanation could be the formation of hydrogen accelerating the polymerization as suggested by our industrial cooperation partner (see also Resconi et al. [60]).

In order to test the hypothesis, additional polymerizations at different temperatures were carried out without hydrogen, whereby gas-phase samples were taken during the reaction at several times and analyzed with a μ-GC (CP-4900, Varian). The resulting chromatogram of the sample is shown in Figure 94 a) together with the pure monomer as well as nitrogen used for catalyst injection. A clear peak at the retention time of hydrogen (0.45 min) could be detected for the sample, whereas the pure gases show no peak indicating indeed the formation of hydrogen. This result therefore confirms that a further reaction must be taken place leading to the formation of hydrogen. Nevertheless, for a quantitative description further experimental studies are necessary, which is not within the scope of this work.

a)                                      b)

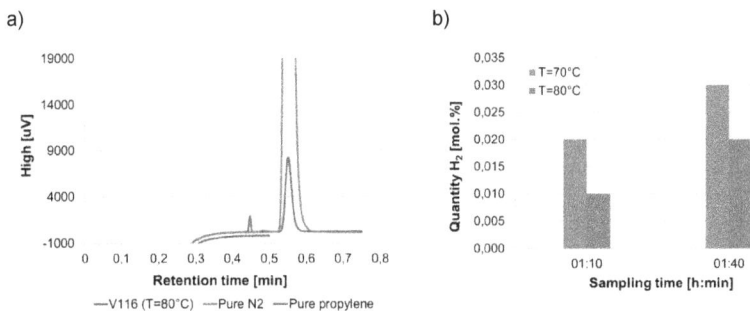

Figure 94: Results bulk polymerization with metallocene catalyst: Polymerization without hydrogen a) Chromatogram of gas-phase sample together with pure monomer and nitrogen; b) Measured hydrogen concentration in the sample taken after different reaction times, (T=70 and 80°C)

### 7.2.3.3 Results weight average molecular weight

The comparison of the simulated and experimentally determined weight average molecular weights is shown in Figure 95. Herein, the weight avg. molecular weights derived at the different $H_2$ concentrations are shown for the investigated temperature range from 55°C to 80°C.

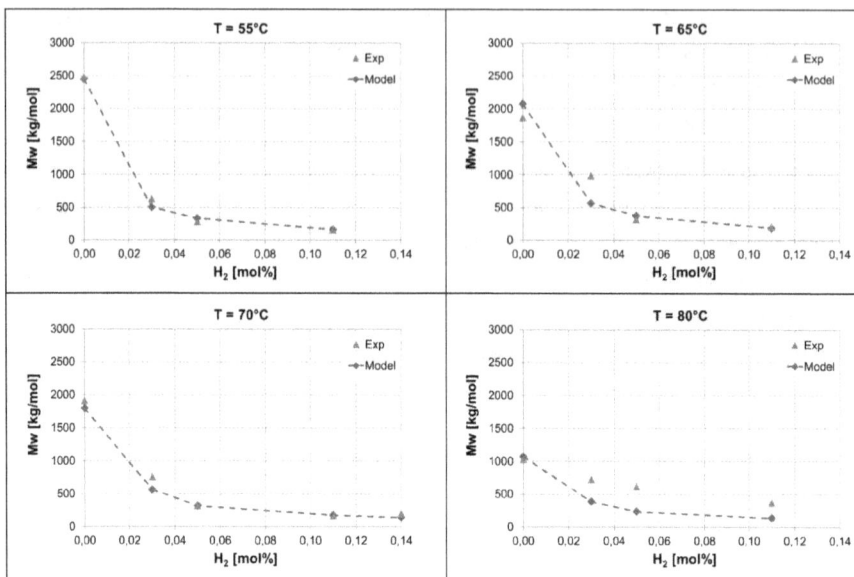

Figure 95: Results kinetic modeling bulk polymerization with metallocene catalyst: Comparison simulated and experimentally derived weight average molecular weights (T=55-80°C, $H_{2,feed}$=0-0.11 mol%)

The weight avg. molecular weights can be predicted by the model in a quantitative good manner from 55°C to 70°C. Only at higher reaction temperature of 80°C, the weight avg. molecular weights are underestimated by the model. A possible reason could be a different real hydrogen concentration in the liquid phase than it was calculated within the model.

## 7.3 Summary kinetic modeling of the bulk polymerization with a supported metallocene catalyst

Aim of the modeling activities was to develop a simplified phenomenological model approach describing the kinetics of the bulk polymerization of propylene with a prepolymerized supported metallocene catalyst at different reaction conditions (different reaction temperatures and $H_2$ concentrations). The simplified kinetic scheme was derived based on the experimental data and comprises the elementary steps of chain initiation with the already activated catalyst, chain propagation and spontaneous deactivation of the growing chains. The influence of hydrogen on the catalyst activity was described based on the dormant site theory and involves the formation of dormant sites and the reactivation of dormant sites by hydrogen as well as spontaneous reactivation. As transfer reactions, chain transfer to hydrogen and spontaneous chain transfer were considered in order to describe the molecular weights.

As metallocene catalysts are single site catalysts, only one sort of active sites was assumed. For the deviation of the mass balances of the involved components, a quasi-homogenous particle model was assumed, wherein mass and heat transfer limitations as well as swelling by monomer were neglected. Temperature dependencies of the kinetic constants were described by means of the Arrhenius equation. For the calculation of the equilibrium monomer concentration at the active sites the Flory-Huggins equation was used. Therein, the interaction parameter was estimated using the Laar-Hildebrand equation. Hydrogen concentration in the polymer particle was considered to be equal to the hydrogen concentration in the liquid phase and it was calculated by Henry's law. Average molecular weights were calculated with the method of moments.

Simulations as well as parameter estimation were carried out with gProms ModelBuilder. For the estimation of the kinetic parameters the experimentally derived activity profiles and molecular weights were fitted. The comparison of simulated and experimentally derived avg. activities as well as avg. molecular weights showed a good agreement within an error range of 10%. The activity profiles can be predicted in an adequate manner. The here developed simplified model approach can be used to quantitatively describe the kinetics of the polymerization reaction within the investigated reaction conditions.

From the comparison of the modeled and experimental data, further phenomena could be observed such as the formation of hydrogen during the polymerization with metallocene catalyst and a possible more pronounced influence of hydrogen at the beginning of the reaction. For further studies it is therefore suggested to investigate the effect of hydrogen at the early polymerization stage as well as the formation of hydrogen in more detail.

# 8 Summary of the work

Aim of this work was to investigate the kinetics of different supported coordination catalysts for the polymerization of propylene under gas-phase as well as under bulk conditions and to develop simplified phenomenological kinetic models describing the polymerizations. A particular focus was to study the effect of prepolymerization on catalyst kinetics.

In the first part of the work, the gas-phase polymerization of propylene with two different 4th generation Ziegler-Natta catalysts was studied. Therein, the effects of different reaction temperatures, pressures and hydrogen concentrations as well as the effect of different catalyst injection conditions (without and with in-situ prepolymerization) on the catalyst activity as well as on polymer characteristics have been investigated. The polymerizations were carried out in a 5 l horizontal stirred tank reactor. Polymerization kinetics were monitored online by feeding of monomer in a semi-batch mode of operation via a pressure control-loop keeping isobaric conditions in the reactor. Polymerizations were carried out in the temperature range from 50°C to 90°C, at pressures between 25 and 30 bars and with hydrogen concentrations between 0 and 0.05 mol/l. Both procedures with and without prepolymerization have been applied. For both catalysts similar trends could be observed.

The average activity linearly increases with increasing pressure due to the increase of the equilibrium monomer concentration at the active sites within the polymer particle. The pressure effect is more pronounced for catalyst B compared to catalyst A due to the higher activity of catalyst B. Both catalysts also showed a similar hydrogen response with respect to average activities as well as molecular weights of the polymer. With increasing hydrogen concentration up to 0.025 mol/l, average activities increased, where an activity plateau was reached. A further increase of hydrogen up to 0.05 mol/l led to no further increase of activity. Hydrogen acts as chain transfer agent and therefore, as expected, molecular weights decreased with increasing hydrogen concentrations. The average activity increased with increasing reaction temperature up to 70°C. For polymerizations without prepolymerization, a further increase above 70°C led to a strong decrease of the average activity. The same behavior was observed for both catalysts, whereas the activity decrease at higher reaction temperatures was more pronounced for the higher active catalyst B. The application of a prepolymerization step significantly improved the activities and particle morphology of both catalysts at higher reaction temperatures. The impact of prepolymerization is catalyst specific and depends on the catalyst activity reached at the polymerization temperature. In general, the higher the catalyst activity is the more critical is the catalyst injection temperature on the final reached average activity.

Based on the experimental investigations, a simplified kinetic model was developed which is able to describe the polymerization at the different reaction conditions including the effect of prepolymerization. Main hypothesis for modeling the effect of prepolymerization is that particle overheating at the beginning of the reaction is the main reason for the lower activities obtained when no prepolymerization step was applied. In the developed model, the elementary steps formation of active sites, chain propagation, chain transfer by hydrogen and spontaneous as well as spontaneous deactivation are considered. The influence of hydrogen on the activity is described based on the dormant site theory. In order to describe the strong temperature effect on activity at high reaction temperatures as well as the different catalyst injection conditions, an additional strong temperature dependent thermal deactivation reaction is considered. As simplification, only one sort of active sites is assumed. For modeling, a quasi-homogeneous particle model is assumed wherein particle growth as well as particle heat transfer are considered. Heat removal is calculated based on the Ranz-Marshall Nusselt-correlation. The temperature dependency of the kinetic constants is described by Arrhenius equations which are related to the particle temperature. Equilibrium monomer concentrations in the polymer particle are calculated according Henry's law and Stern equation. Average molecular weights are calculated using the method of moments.

The rate constants as model parameters were determined by parameter estimation to the experimental results. Herein, both Ziegler-Natta catalysts can be described with the same set of kinetic constants; the only difference in the model being the amount of active component for each catalyst. With the developed simplified kinetic model, simulated average activities, activity profiles as well as weight average molecular weights showed a relative good agreement with the experimentally derived data for both catalysts at the different reaction conditions. Also the influence of the different injection conditions could be described in a quantitative good manner.

In the second part of this work, the bulk polymerization of propylene with a supported metallocene catalyst has been studied. The influences of reaction temperature, hydrogen concentration and prepolymerization degree on catalyst activity as well as polymer characteristics were investigated.

In order to study the kinetics of bulk phase polymerization, a special reaction calorimeter was used which measures the heat flow generated by the exothermal polymerization reaction through heat conductivity in the reactor base. The bulk polymerizations were carried out in the temperature range from 55°C to 80°C and with hydrogen feed concentrations from 0 to 0.14 mol%. Different polymerization procedures were developed: Polymerization with in-situ prepolymerization and polymerization with a prepolymerized catalyst, which was produced in a separate prepolymerization beforehand. With both procedures same average activities as well as MFR values have been obtained with a high reproducibility. The final kinetic measurements were carried out with the prepolymerized catalyst as the kinetic information are earlier accessible and prepolymerization conditions are more defined for the external procedure. The degree of prepolymerization has an influence on the final activity reached at main polymerization conditions. For the studied catalyst, a maximum activity was reached in the main polymerization for a prepolymerization degree of 160 $mg_{PP}/mg_{Cat}$, which was used for final kinetic measurements. The average activities increased with increasing temperature from 55°C to 70°C. Higher reaction temperatures of 80°C led to a decrease in activity. Hydrogen accelerates the polymerization reaction. With increasing hydrogen feed concentration up to 0.05 mol% the activity increased. At higher hydrogen feed concentrations up to 0.14 mol%, no further increase of activity was observed. Without hydrogen, low activities as well as a slow reaction start were observed. As expected, with increasing hydrogen concentration, the molecular weights decreased.

Also for the bulk-phase polymerization with the studied prepolymerized supported metallocene catalyst, a simplified phenomenological kinetic model was developed. Therein, chain initiation of the activated catalyst, propagation, transfer with hydrogen and spontaneous chain transfer and spontaneous deactivation are considered. The influence of hydrogen on the activity is described with the dormant site theory. For balancing, a quasi-homogenous particle model is applied wherein both mass transfer limitations as well as swelling by monomer are neglected. No heat transfer limitations are assumed since only prepolymerized catalyst was studied and in bulk polymerization, in general, heat removal is much better compared to gas-phase polymerization. Partitioning of propylene and hydrogen between gas-phase and liquid phase is considered. The equilibrium monomer concentration is calculated with the Flory-Huggins equation. The interaction parameter is estimated using the Laar-Hildebrand equation. Hydrogen concentration in the polymer particle is considered to be equal to the hydrogen concentration in the liquid phase and is calculated by Henry's law. Temperature dependencies of the kinetic constants are described by means of the Arrhenius equation. Average molecular weights are calculated with the method of moments.

With the developed model and the estimated kinetic parameters, the average activities and molecular weights can be calculated within a 10 % error range to the experimental data. The activity profiles can be predicted in an adequate manner at the different reaction temperatures as well as with different hydrogen concentrations. Differences between model and experimental data were found for polymerizations without hydrogen. A possible explanation could be the formation of small amounts of hydrogen with the used metallocene catalyst which accelerates the polymerization.

# 9 Appendix

## 9.1 Gas-phase polymerization of propylene with different Ziegler-Natta catalysts

### 9.1.1 Experimental results gas-phase polymerization of propylene with Ziegler-Natta catalysts A and B

Table 27: Reproducibility of gas-phase polymerization reactions: Average activities and MFR values (T=80°C, p=27.5 bar, $H_2$=0.025 mol/l, without prepolymerization, cat A)

| Exp. | $m_{Cat}$ | $H_2$ | Avg. activity | MFR | $M_w$ |
|------|------|------|------|------|------|
|      | mg | mol/l | $kg_{PP}/(g_{Cat} \cdot h)$ | g/10 min | $10^3 \cdot$g/mol |
| V27 | 19.0 | 0.240 | 9.0 | 75.2 | 183.6 |
| V31 | 17.7 | 0.245 | 8.7 | 70.8 | 186.4 |
| V38 | 16.8 | 0.250 | 8.5 | 79.4 | 181.2 |

Table 28: Results gas-phase polymerization: Influence of reaction pressure ($H_2$=0.025 mol/l, no prepoly)

| Exp. | Catalyst | T | p | $c_M{}^*$ | A | $A/c_M{}^*$ | $M_w$ |
|------|------|------|------|------|------|------|------|
|      |      | °C | bar | mol/l | $kg_{PP}/(g_{Cat} \cdot h)$ | $(kg_{PP}/(g_{Cat} \cdot h)/(g/l)$ | $10^3 \cdot$g/mol |
| V30 | A | 80 | 25 | 1.470 | 7.88 | 0.1274 | 176.9 |
| V31 | A | 80 | 27.5 | 1.617 | 8.68 | 0.1276 | 186.4 |
| V29 | A | 80 | 30 | 1.763 | 9.11 | 0.1227 | 190.9 |
| V21 | B | 70 | 25 | 1.720 | 15.49 | 0.2140 | 254.3 |
| V7 | B | 70 | 27.5 | 1.892 | 19.18 | 0.2409 | 261.0 |
| V22 | B | 70 | 30 | 2.064 | 25.30 | 0.2914 | 257.9 |

$c_M{}^*$: equilibrium monomer concentration in polymer particle

Table 29: Results gas-phase polymerization: Influence of reaction temperature ($H_2$=0.025 mol/l, no prepoly)

| Exp. | Catalyst | T | p | $c_M{}^*$ | A | $A/c_M{}^*$ | $M_w$ |
|------|------|------|------|------|------|------|------|
|      |      | °C | bar | mol/l | $kg_{PP}/(g_{Cat} \cdot h)$ | $(kg_{PP}/(g_{Cat} \cdot h)/(g/l)$ | $10^3 \cdot$g/mol |
| V55 | A | 50 | 16.4 | 1.616 | 4.0 | 0.060 | 232.1 |
| V54 | A | 60 | 19.8 | 1.617 | 8.0 | 0.117 | 248.2 |
| V34 | A | 70 | 27.5 | 1.892 | 13.5 | 0.170 | 265.4 |
| V38 | A | 80 | 27.5 | 1.616 | 8.49 | 0.125 | 181.1 |
| V35 | A | 90 | 27.5 | 1.399 | 5.18 | 0.088 | 141.1 |
| V08 | B | 50 | 16.4 | 1.616 | 9.6 | 0.142 | 230.0 |
| V05 | B | 60 | 19.8 | 1.617 | 12.7 | 0.186 | 233.9 |
| V27 | B | 70 | 27.5 | 1.892 | 18.6 | 0.234 | 261.0 |
| V06 | B | 80 | 27.5 | 1.616 | 8.4 | 0.124 | 161.5 |
| V04 | B | 90 | 27.5 | 1.399 | 3.4 | 0.058 | 120.5 |

$c_M{}^*$: equilibrium monomer concentration in polymer particle

**Table 30: Results gas-phase polymerization: Influence of injection conditions ($H_2$=0.025 mol/l, with prepoly)**

| Exp. | Catalyst | T | p | $c_M$* | A | $A/c_M$* | $M_w$ |
|------|----------|---|---|--------|---|----------|-------|
| | | °C | bar | mol/l | $kg_{PP}/(g_{Cat}·h)$ | $(kg_{PP}/(g_{Cat}·h)/(g/l)$ | $10^3·g/mol$ |
| V50 | A | 70 | 27.5 | 1.892 | 12.9 | 0.162 | 274.6 |
| V58 | A | 80 | 27.5 | 1.616 | 14.1 | 0.207 | 245.3 |
| V49 | A | 90 | 27.5 | 1.399 | 10.0 | 0.170 | 202.1 |
| V25 | B | 60 | 19.8 | 1.617 | 16.0 | 0.235 | 242.7 |
| V23 | B | 70 | 27.5 | 1.892 | 30.2 | 0.379 | 256.8 |
| V26 | B | 80 | 27.5 | 1.616 | 31.0 | 0.456 | 229.3 |
| V24 | B | 90 | 27.5 | 1.399 | 15.7 | 0.267 | 181.8 |

$c_M$*: equilibrium monomer concentration in polymer particle

## 9.1.2  Overview polymer characterization

### 9.1.2.1  DSC measurements

Figure 96 shows an example of a DSC curve from a polypropylene sample together with the applied temperature program. Therein, the measured heat flow is related to the sample weight. For the determination of the crystallinity of the powder sample, the specific heat enthalpy was derived from integrating the first melting peak of the heat flow curve and related to the specific enthalpy of an ideal crystallite.

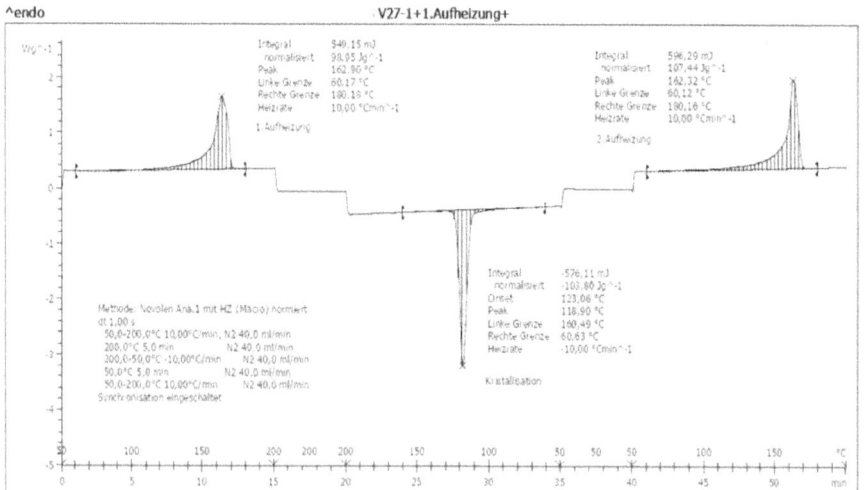

Figure 96: DSC curve of a polypropylene sample (powder) produced in gas-phase polymerization

Table 31: Results analytics gas-phase polymerization: DSC measurements – Analysis of the first melting peak and calculated crystallinity

| | | | Catalyst A | | | | Catalyst B | | | |
|---|---|---|---|---|---|---|---|---|---|---|
| T | $H_2$ | Prepoly | Exp. | $\Delta H_m$ | $T_m$ | K | Exp. | $\Delta H_m$ | $T_m$ | K |
| °C | mol/l | | | J/g | °C | % | | J/g | °C | % |
| 50 | 0.025 | No | - | - | - | - | 8 | 112.4 | 163.1 | 54.3 |
| 70 | 0 | No | 36 | 103.9 | 164.9 | 50.2 | 9 | 102.4 | 165.6 | 49.5 |
| 70 | 0.025 | No | 34 | 108.3 | 162.5 | 52.3 | 27 | 106.3 | 163.2 | 51.3 |
| 70 | 0.025 | Yes | 50 | 108.8 | 162.3 | 52.5 | 23 | 105.7 | 162.1 | 51.1 |
| 70 | 0.05 | No | 42 | 112.5 | 163.9 | 54.3 | 16 | 105.7 | 161.9 | 51.1 |
| 90 | 0.025 | No | 35 | 101.9 | 156.5 | 49.2 | 4 | 102.5 | 159.8 | 49.5 |
| 90 | 0.025 | Yes | 49 | 107.4 | 161.2 | 51.9 | - | - | - | - |

## 9.1.2.2 Results particle size and particle size distribution measurements

Table 32: Results analytics gas-phase polymerization: PS and PSD determined with Malvern and by sieving analysis: Influence of temperature and $H_2$ concentration on particle size (cat A and B, no prepoly)

| | | | Malvern | | Sieving | | | | | | |
|---|---|---|---|---|---|---|---|---|---|---|---|
| Exp. | T | $H_2$ | D50 (< 2000 µm) | >2000 µm | <400 µm | 400 - 1000 µm | >1000 µm | Loss | BD | Activity |
| | °C | mol/l | µm | wt% | wt% | wt% | wt% | wt% | g/ml | kg$_{PP}$/g$_{Cat}$/h |
| Cat A | | | | | | | | | | |
| V55 | 50 | 0.025 | 835 | 2 | - | - | - | - | 0.390 | 4.04 |
| V34 | 70 | 0.025 | 887 | 18.1 | 4.20 | 14.94 | 80.32 | 0.55 | 0.356 | 13.54 |
| V40 | 80 | 0 | 832 | 10.5 | 2.46 | 34.09 | 63.00 | 0.45 | - | 3.96 |
| V38 | 80 | 0.025 | 865 | 4.2 | 4.85 | 21.80 | 71.76 | 1.60 | 0.378 | 8.49 |
| V33 | 80 | 0.050 | 868 | 1.3 | 7.68 | 27.22 | 64.83 | 0.27 | 0.373 | 8.86 |
| V35 | 90 | 0.025 | 801 | 0.5 | 7.87 | 42.99 | 48.26 | 0.88 | 0.315 | 5.13 |
| Cat B | | | | | | | | | | |
| V8 | 50 | 0.025 | 948 | 0 | | | | | 0.404 | 9.62 |
| V27 | 70 | 0.025 | 885 | 22 | | | | | 0.330 | 18.59 |
| V6 | 80 | 0.025 | 958 | 16 | | | | | 0.299 | 8.42 |
| V17 | 80 | 0.050 | 835 | 2 | | | | | 0.289 | 8.12 |
| V4 | 90 | 0.025 | 666 | 14 | | | | | - | 3.43 |

## 9.1.2.3 Polymer density measurements

The polymer density was measured with helium pycnometry (Pycnomatic ATC, Porotec GmbH) at the Department of Industrial Chemistry at the Martin-Luther-University Halle-Wittenberg. The results are summarized in Table 33. Similar polymer densities were measured for the polymer powders produced at different reaction conditions as well as with the different catalysts. Therefore, an average polymer density of 0.905 g/ml can was used for calculations for both ZN catalysts.

Table 33: Results analytics gas-phase polymerization: Polymer density of polymer produced with catalyst A and B at different reaction conditions

| Catalyst | T [°C] | $H_2$ [mol/l] | Prepoly | Polymer density [g/ml] |
|----------|--------|---------------|---------|------------------------|
| B | 50 | 0.025 | No | 0.896 |
| B | 70 | 0 | No | 0.903 |
| B | 70 | 0.025 | No | 0.913 |
| B | 70 | 0.025 | Yes | 0.904 |
| A | 70 | 0.025 | No | 0.911 |

## 9.1.3 Prepolymerization experiment

The prepolymerization was carried out with catalyst B according procedure 2 (see chapter 4.3.3) but directly stopped after the prepolymerization phase. Because temperature and pressure were not constant during prepolymerization, no kinetic data could be recorded. Table 34 summarizes the reaction conditions and results of the prepolymerization experiment (V38) together with the polymerization experiments with and without prepolymerization step (V26 and V6, respectively) at same reaction conditions.

Table 34: Results gas-phase polymerization catalyst B: Prepolymerization experiment compared with polymerizations with and without prepolymerization step (T=80°C, $H_2$=0.025 mol/l)

| Exp. | Pre-poly | m Cat | Time | T Cat-inj. | T Polym. | p Cat-inj. | p Polym. | m PP | Yield | Avg. activity |
|------|----------|-------|------|------------|----------|------------|----------|------|-------|---------------|
| | | mg | h | °C | °C | bar | bar | g | $kg_{PP}/g_{Cat}$ | $kg_{PP}/g_{Cat}/h$ |
| V38 | x | 14.7 | 00:15 | 40 | 80* | 11.7 | 25.0* | 114.9 | 7.8 | 31.3 |
| V26 | x | 8.4 | 01:15 | 39 | 80 | 13.4 | 27.5 | 325.9 | 38.8 | 31.0 |
| V6 | - | 10.3 | 01:00 | 79 | 80 | 26.6 | 27.5 | 86.7 | 8.4 | 8.4 |

*T and p after 15 min of prepolymerization

## 9.2 Bulk polymerization of propylene with a supported metallocene catalyst

### 9.2.1 Results bulk polymerization and polymer characterization

Table 35: Results bulk polymerization and analytics: Average activities, MFR and results GPC measurements

| Exp. | T | $H_2$ | $H_2$ (real) | Avg. activity | MFR | GPC measurements $M_n$ | $M_w$ | $M_w/M_n$ |
|------|-----|-----|--------|-------|------|--------|---------|-------|
|      | °C  | mg  | mol%   | $kg_{PP}/g_{Cat}/h$ | g/10 min | kg/mol | kg/mol | - |
| 90  | 55 | 0 | 0     | 3.07  | <0.1 | 164.44 | 2,481.49 | 15.09 |
| 103 |    | 1 | 0.033 | 11.9  | 0.6  | 171.23 | 627.52   | 3.66  |
| 86  |    | 2 | 0.060 | 16.9  | 11.9 | 83.45  | 276.67   | 3.32  |
| 91  |    | 4 | 0.112 | 16.0  | 75   | 46.54  | 151.71   | 3.26  |
| 95  | 65 | 0 | 0.000 | 3.9   | <0.1 | 192.06 | 1863.61  | 9.70  |
| 109 |    | 1 | 0.029 | 15.7  | 0.1  | 276.17 | 984.85   | 3.57  |
| 94  |    | 2 | 0.048 | 21.6  | 8.8  | 108.28 | 315.65   | 2.92  |
| 96  |    | 4 | 0.112 | 23.0  | 29.1 | 69.21  | 205.88   | 2.97  |
| 88  | 70 | 0 | 0.000 | 5.9   | <0.1 | 265.39 | 1,659.81 | 6.25  |
| 97  |    | 1 | 0.027 | 17.7  | 0.4  | 214.69 | 751.98   | 3.50  |
| 87  |    | 2 | 0.055 | 24.9  | 5.3  | 112.04 | 316.87   | 2.83  |
| 89  |    | 4 | 0.113 | 24.1  | 54.6 | 58.84  | 162,23   | 2.76  |
| 98  |    | 5 | 0.138 | 26.9  | 8.6  | 66.35  | 190.23   | 2.87  |
| 101 | 80 | 0 | 0.000 | 4.9   | <0.1 | 166.56 | 1,030.14 | 6.18  |
| 104 |    | 1 | 0.030 | 18.7  | 0.2  | 211.19 | 721.97   | 3.42  |
| 100 |    | 2 | 0.060 | 19.3  | 0.4  | 173.61 | 616.67   | 3.55  |
| 102 |    | 4 | 0.111 | 19.7  | 4.1  | 119.98 | 367.32   | 3.06  |

Notice: Mn and Mw for polymerizations without $H_2$ inconclusive, due to difficulties in integration method.

Table 36: Results bulk polymerization and analytics: Crystallinity, porosity and densities

| Exp. | T | $H_2$ | $H_2$ (real) | Crystallinity | Porosity | Polymer density | Bulk density |
|------|-----|-----|--------|------|------|-------|-------|
|      | °C  | mg  | mol%   | %    | %    | g/ml  | g/ml  |
| 86  | 55 | 2 | 0.060 | 46.23 | 21.2 | 0.881 | 0.432 |
| 94  | 65 | 2 | 0.048 | 46.38 | 19.9 | 0.888 | 0.433 |
| 88  | 70 | 0 | 0.000 | 45.36 | 16.5 | 0.886 | 0.433 |
| 97  |    | 1 | 0.027 | 45.41 | 14.2 | -     | 0.453 |
| 87  |    | 2 | 0.055 | 46.71 | 18.3 | 0.899 | 0.429 |
| 89  |    | 4 | 0.113 | 48.89 | 18.0 | -     | 0.420 |
| 98  |    | 5 | 0.138 | 47.78 | 16.4 | 0.888 | 0.437 |
| 100 | 80 | 2 | 0.060 | -     | 12.2 | 0.872 | 0.405 |
| 102 |    | 4 | 0.111 | 45.75 | -    | -     | -     |

### 9.2.2 Derivation of moment equations for the calculation of average molecular weights for the bulk polymerization with metallocene catalyst

Average molecular weights are calculated according to the method of moments analogues as described in chapter 5.1.7. The differential equations for the zeroth, first and second moment of the three polymer species were derived from the developed kinetic scheme for the bulk polymerization (Figure 88, page 126). Following set of differential equations is achieved:

*Growing chains:*

$$\frac{dQ0}{dt} = k_i \cdot n_{Zr*} \cdot c_{C3} - \left(k_{dorm,f} \cdot c_{C3} + k_{des}\right) \cdot Q0 + \left(k_{dorm,r1} \cdot c_{H2}^L + k_{dorm,r2}\right) \cdot R0 \tag{9.1}$$

$$\frac{dQ1}{dt} = k_i \cdot n_{Zr*} \cdot c_{C3} + k_p \cdot c_{C3} \cdot Q0 - \left(k_{dorm,f} \cdot c_{C3} + k_{des}\right) \cdot Q1$$
$$+ \left(k_{dorm,r1} \cdot c_{H2}^L + k_{dorm,r2}\right) \cdot R1 + \left(k_{tr,H2} \cdot c_{H2}^L + k_{tr,\beta-H}\right) \cdot (Q0 - Q1) \tag{9.2}$$

$$\frac{dQ2}{dt} = k_i \cdot n_{Zr*} \cdot c_{C3} + k_p \cdot c_{C3} \cdot (2 \cdot Q1 + Q0) - \left(k_{dorm,f} \cdot c_{C3} + k_{des}\right) \cdot Q2$$
$$+ \left(k_{dorm,r1} \cdot c_{H2}^L + k_{dorm,r2}\right) \cdot R2 + \left(k_{tr,H2} \cdot c_{H2}^L + k_{tr,\beta-H}\right) \cdot (Q0 - Q2) \tag{9.3}$$

*Dormant chains:*

$$\frac{dR0}{dt} = k_{dorm,f} \cdot c_{C3} \cdot Q0 - \left(k_{dorm,r1} \cdot c_{H2}^L + k_{dorm,r2}\right) \cdot R0 \tag{9.4}$$

$$\frac{dR1}{dt} = k_{dorm,f} \cdot c_{C3} \cdot Q1 - \left(k_{dorm,r1} \cdot c_{H2}^L + k_{dorm,r2}\right) \cdot R1 \tag{9.5}$$

$$\frac{dR2}{dt} = k_{dorm,f} \cdot c_{C3} \cdot Q2 - \left(k_{dorm,r1} \cdot c_{H2}^L + k_{dorm,r2}\right) \cdot R2 \tag{9.6}$$

*Dead chains:*

$$\frac{dD0}{dt} = \left(k_{tr,H2} \cdot c_{H2}^L + k_{tr,\beta-H} + k_{des}\right) \cdot Q0 \tag{9.7}$$

$$\frac{dD1}{dt} = \left(k_{tr,H2} \cdot c_{H2}^L + k_{tr,\beta-H} + k_{des}\right) \cdot Q1 \tag{9.8}$$

$$\frac{dD2}{dt} = \left(k_{tr,H2} \cdot c_{H2}^L + k_{tr,\beta-H} + k_{des}\right) \cdot Q0 \tag{9.9}$$

Weight average molecular weight $M_w$, number average molecular weight $M_n$ and PDI is calculated from the zeroth, first and second moment according the equations (5.49) to (5.51) given in chapter 5.1.7.

# References

[1] Plastics Europe (2015): Plastics – the Facts 2015. An analysis of European plastics production, demand and waste data. Edited by Plastics Europe. Belgium. Available online at http://www.plasticseurope.org/Document/plastics---the-facts-2015.aspx?Page=DOCUMENT&FolID=2.

[2] Ceresana (2014): Market Study: Polypropylene. Edited by Ceresana. Germany. Available online at http://www.ceresana.com/en/market-studies/plastics/polypropylene/polypropylene-market-share-capacity-demand-supply-forecast-innovation-application-growth-production-size-industry.html.

[3] Kaiser, W. (2011): Kunststoffchemie für Ingenieure. Von der Synthese bis zur Anwendung. 3. edition. München: Hanser. ISBN: 978-3-446-43047-1.

[4] Doshev, P. *et al.* (2013): Marktentwicklung Polypropylen (PP). In *Kunststoffe* 10, pp. 48–54. Available online at https://www.kunststoffe.de/themen/basics/standardthermoplaste/ polypropylen-pp/artikel/marktentwicklung-polypropylen-pp-714988.

[5] Malpass, D. B.; Band, E. I. (2012): Introduction to industrial polypropylene. Properties, catalysts, processes. Online edition. Hoboken, N.J, Salem, Mass: John Wiley & Sons. ISBN: 9781118062760.

[6] Moore, E. P. (Ed.) (1996): Polypropylene handbook. Polymerization, characterization, properties, processing, applications. München [etc.], Cincinnati, Ohio: Hanser; Hanser-Gardner. ISBN: 978-3-446-18176-2.

[7] Soares, J. B. P.; McKenna, T. F. (2012): Polyolefin reaction engineering. Weinheim, Chichester: Wiley-VCH. ISBN: 978-3-527-31710-3.

[8] Abts, Georg (2014): "Kunststoff-Wissen für Einsteiger", 3. updated and expanded edition, München: Carl Hanser Verlag. ISBN: 978-3-446-45041-7.

[9] Wiley-VCH (Ed.) (2016): Ullmann's Polymers and Plastics: Products and Processes. Volume 2. Part 2: Organic Polymers. Polypropylene. By: M. Gahleitner, C. Paulik. 1. edition. Weinheim, Germany: Wiley-VCH Verlag GmbH & Co. KGaA. ISBN: 978-3-527-33823-8. pp. 937-980.

[10] Kissin, Y. V.; Rishina, L.A.; Vizen, E. I. (2002): Hydrogen effects in propylene polymerization reactions with titanium-based Ziegler-Natta catalysts. II. Mechanism of the chain-transfer reaction. In *J. Polym. Sci. A Polym. Chem.* 40, pp. 1899–1911.

[11] Bartke, M. (2011): Koordinative Polymerisation von Olefinen. Kinetik und Verfahren. DECHEMA-Kurs "Polymerisationstechnik". Hamburg, 26.-30.09.2011.

[12] Kim, J. D.; Soares, J. B. P. (2000): Copolymerization of ethylene and alpha-olefins with combined metallocene catalysts. III. Production of polyolefins with controlled microstructures. In *J. Polym. Sci. A Polym. Chem.* 38 (9), pp. 1427–1432.

[13] Blackmon, K. P.; Thorman, J. L.; Malbari, S. A.; Wallace, M. (2010): Succinate-containing polymerization catalyst system using n-butylmethyldimethoxysilane for preparation of polypropylene film grade resins. Applied for by Fina Technology, Inc. on 18.12.2007. Patent no. US7851578 B2.

[14] Kaminsky, W.; Laban, A. (2001): Metallocene catalysis. In *Applied Catalysis A: General* 222 (1-2), pp. 47–61.

[15] Asua, J. M. (Ed.) (2007): Polymer reaction engineering. Chapter 2: Coordination Polymerization. By: J. B. P. Soares, T. F. McKenna, C. P. Cheng. Oxford, Ames, Iowa: Blackwell Pub. ISBN: 978-1-4051-4442-1. pp. 29-117.

[16] The Dow Chemical Company (2013): Industry Leading UNIPOL™ PP Process Technology Selected by QP/QAPCO for Polypropylene Facility. Midland, Michigan. Available online at http://www.dow.com/en-us/news/press-releases/Industry%20Leading%20UNIPOL%20PP%20Process%20Technology%20Selected%20by%20QPQAPCO%20for%20Polypropylene%20Facility.

[17] Tait, P. J.T. (1989): Monoalkene Polymerization: Ziegler–Natta and Transition Metal Catalysts. In *Comprehensive Polymer Science and Supplements* 4, pp. 1–25.

[18] CB&I (2017): Novolen® Gas-Phase Process. Available online at http://www.cbi.com/What-We-Do/Technology/Petrochemicals/Olefins/Polypropylene-Production/Novolen-Gas-Phase-Process.

[19] CB&I (2016): Polypropylene. Available online at http://www.cbi.com/getattachment/a50c26ce-df7a-4268-8ee1-f806af21648c/Polypropylene.aspx.

[20] LyondellBasell Industries Holdings B.V. (2016): Licensed Polyolefin Technologies and Services. Spheripol. Available online at https://www.lyondellbasell.com/globalassets/products-technology/technology/spheripol-brochure.pdf.

[21] LyondellBasell Industries Holdings B.V. (2016): Licensed Polyolefin Technologies and Services. Spherizone. Available online at https://www.lyondellbasell.com/globalassets/products-technology/technology/spherizone-brochure.pdf.

[22] Mitsui Chemicals, Inc. (2017): Mitsui Polypropylene Technology. Available online at http://www.mitsuichem.com/techno/license/pdf/pp_process.pdf.

[23] Borealis AG (2017): Technologies. Borstar Technology. Available online at http://www.borealisgroup.com/en/company/innovation/technologies/.

[24] Denifl, P.; Leinonen, T. (2001): Preparation of olefin polymerisation catalyst component. Applied for by Borealis Technology Oy on 20.6.2001. Patent no. EP1273595 A1.

[25] Denifl, P.; Preat, E. Van; Bartke, M.; Oksman, M.; Mustonen, M.; Garoff, T.; Pesonen, K. (2002): Production of olefin polymerisation catalysts. Applied for by Borealis Technology Oy on 18.12.2002. Patent no. WO2003051934 A2.

[26] McKenna, T. F.; Soares, J. B. P. (2001): Single particle modelling for olefin polymerization on supported catalysts: A review and proposals for future developments. In *Chem. Eng. Sci.* 56, pp. 3931–3949.

[27] Keii, T.; Doi, Y.; Suzuki, E.; Tamura, M.; Murata, M.; Soga, K. (1984): Propene polymerization with a magnesium chloride-supported Ziegler catalyst, 2†. Molecular weight distribution. In *Makromol. Chem.* 185 (8), pp. 1537–1557.

[28] Soares, J. B. P.; Hamielec, A. E. (1996): Kinetics of propylene polymerization with a non-supported heterogeneous Ziegler-Natta catalyst—effect of hydrogen on rate of polymerization, stereoregularity, and molecular weight distribution. In *Polymer* 37 (20), pp. 4607–4614.

[29] Ziegler, K.; Breil, H.; Holzkamp, E.; Martin, H. (1960): Verfahren zur Herstellung von hochmolekularen Polyäthylenen. Applied for by Ziegler, K. on 17.11.1953. Patent no. DE973626 C.

[30] Natta,G., Pino, P., Mazzanti, G. (1954):

    a) Italian Patent, Patent no. 535712, 8.6.1954.

    b) Italian Patent, Patent no. 537425, 27.7.1954.

    c) Italian Patent, Patent no. 526101, 4.12.1954.

[31] Tornqvist, E.; Langer, A.W., Jr (1958): Polymerization catalyst. Applied for by Exxon Research Engineering Co on 27.6.1958. Patent no. US3032510 A.

[32] Hermans, J. P.; Henrioulle, P. (1971): Process for the preparation of a ziegler-natta type catalyst. Applied for by Solvay & Cie on 24.3.1971.US Patent no. US3769233 A.

[33] Hermans, J. P.; Henrioulle, P. (1972): Applied for by Solvay & Cie. on 17.3.1972. German Patent no. DE2213086 C2.

[34] Mayr, A.; Galli, P.; Susa, E.; Drusco, G.D.; Giachetti, E. (1968): Polymerization of olefins. Applied for by Montedison. British Patent no. 1,286,867.

[35] Giannini, U.; Cassata, A.; Longi, P.; Mazzoch, R. (1972). Applied for by Montedison. Belgian Patent no. 785,332 and 785,334.

[36] Parodi, S.; Nocci, R.; Giannini, U.; Barbè, P.C.; Scata, U. (1982): Components and catalysts for the polymerization of olefins. Applied for by Montedison. European Patent no. 45,977.

[37] Albizzati, E.; Barbe, P. C.; Noristi, L.; Scordamaglia, R.; Barino, L.; Giannini, U.; Morini, G. (1990): Components and catalysts for the polymerization of olefins. Applied for by Himont Incorporated. US Patent no. US4971937 A.

[38] Brintzinger, H. H.; Fischer, D.; Mülhaupt, R.; Rieger, B.; Waymouth, R. M. (1995): Stereospecific Olefin Polymerization with Chiral Metallocene Catalysts. In *Angew. Chem. Int. Ed. Engl.* 34 (11), pp. 1143–1170.

[39]  Resconi, L.; Cavallo, L.; Fait, A.; Piemontesi, F. (2000): Selectivity in Propene Polymerization with Metallocene Catalysts. In *Chem. Rev.* 100 (4), pp. 1253–1346.

[40]  Kaminsky, Walter (2004): The discovery of metallocene catalysts and their present state of the art. In *J. Polym. Sci. A Polym. Chem.* 42 (16), pp. 3911–3921.

[41]  Natta, G.; Pino, P.; Mazzanati, G.; Giannini; U. (1957): In *J. Am. Chem. Soc.* 79, pp. 2975.

[42]  Breslow, D.S.; Newburg, N.R. (1957): In *J. Am. Chem. Soc.* 79, pp. 5072.

[43]  Sinn, H.; Kaminsky, W.; Vollmer, H.-J.; Woldt, R (1980): "Living Polymers" on Polymerization with Extremely Productive Ziegler Catalysts. In *Angew. Chem. Int. Ed. Engl.* 19 (5), pp. 390–392.

[44]  Schnutenhaus, H.; Brintzinger, H. H. (1979). In *Angew. Chem. Int. Ed. Engl.* 18, pp. 777–778.

[45]  Wild, F. R. W. P.; Zsolnai, L.; Huttner, G.; Brintzinger, H. H. (1982): ansa-Metallocene derivatives. IV. Synthesis and molecular structures of chiral ansa-titanocene derivatives with bridged tetrahydroindenyl ligands. In *Organomet. Chem.* 232, pp. 233–247.

[46]  Kaminsky, W.; Külper, K.; Brintzinger, H. H.; Wild, F. R. W. P. (1985): Polymerization of propene and butene with a chiral zirconocene and methylalumoxane as cocatalyst. In *Angew. Chem. Int. Ed. Engl.* 24, pp. 507–508.

[47]  Hlatky, G. G. (2000): Heterogeneous Single-Site Catalysts for Olefin Polymerization. In *Chem. Rev.* 100 (4), pp. 1347–1376.

[48]  Fink, G.; Steinmetz, B.; Zechlin, J.; Przybyla, C.; Tesche, B. (2000): Propene Polymerization with Silica-Supported Metallocene/MAO Catalysts. In *Chem. Rev.* 100 (4), pp. 1377–1390.

[49]  Hlatky, G. G. (1999): Metallocene catalysts for olefin polymerization: Annual review for 1996. In *Coordination Chemistry Reviews* 181 (1), pp. 243–296.

[50]  Kaminsky, Walter (1998): Highly active metallocene catalysts for olefin polymerization. In *J. Chem. Soc., Dalton Trans.* (9), pp. 1413–1418.

[51]  Ray, W. H. (1986): Modelling of polymerization phenomena. In *Ber. Bunsen Ges. Phys. Chem.* 90, pp. 947–955.

[52]  Ray, W. H. (1991): Modelling of addition polymerization processes - Free radical, ionic, group transfer, and ziegler-natta kinetics. In *Can. J. Chem. Eng.* 69 (3), pp. 626–629.

[53]  Kiparissides, C. (1996): Polymerization reactor modeling: A review of recent developments and future directions. In *Chem. Eng. Sci.* 51 (10), pp. 1637–1659.

[54]  Dubé, M. A.; Soares, J. B. P.; Hamielec, A. E. (1997): Mathematical Modeling of Multicomponent Chain-GrowthPolymerizations in Batch, Semibatch, and Continuous Reactors: A Review. In *Ind. Eng. Chem. Res.* 36 (4), pp. 966–1015.

[55]  Kissin, Y. V. (2012): Active centers in Ziegler–Natta catalysts: Formation kinetics and structure. In *J. Catal.* 292, pp. 188–200.

[56]  Kissin, Jurij V. (2008): Alkene polymerization reactions with transition metal catalysts. 1. ed. Amsterdam: Elsevier (Studies in surface science and catalysis, 173). ISBN:9780444532152. Chapter 6.

[57]  Dusseault, J. J. A.; Hsu, C. C. (1993): MgCl2 -Supported Ziegler-Natta Catalysts for Olefin Polymerization: Basic Structure, Mechanism, and Kinetic Behavior. In *J. Macromol. Sci., Polym. Rev.* 33 (2), pp. 103–145.

[58]  Cossee, P. (1964): Ziegler-Natta catalysis I. Mechanism of polymerization of α-olefins with Ziegler-Natta catalysts. In *J. Catal.* 3 (1), pp. 80–88.

[59]  Arlman, E. J.; Cossee, P. (1964): Ziegler-Natta catalysis III. Stereospecific polymerization of propene with the catalyst system TiCl3/AlEt3. *J. Catal.* 3 (1), pp. 99–104.

[60]  Resconi, L.; Camurati, I.; Sudmeijer, O. (1999): Chain transfer reactions in propylene polymerization with zirconocene catalysts. In *Top. Catal.* 7 (1-4), pp. 145–163.

[61]  Choi, K.-Y.; Ray, W. H. (1985): Polymerization of olefins through heterogeneous catalysis. II. Kinetics of gas phase propylene polymerization with Ziegler-Natta catalysts. In *J. Appl. Polym. Sci.* 30, pp. 1065–1081.

[62] Pater, J. T. M.; Weickert, G.; van Swaaij, W. P. M. (2002): Polymerization of liquid propylene with a 4th generation Ziegler–Natta catalyst—influence of temperature, hydrogen and monomer concentration and prepolymerization method on polymerization kinetics. In *Chem. Eng. Sci.* 57, pp. 3461–3477.

[63] Samson, J.J.C.; van Middelkoop, B.; Weickert, G.; Westerterp, K. R. (1999): Gas-phase polymerization of propylene with a highly active ziegler-natta catalyst. In *AlChE J.* 45 (7), pp. 1548 – 1558.

[64] Meier, G. B.; Weickert, G.; van Swaaij, Wim P. M. (2001): Gas-phase polymerization of propylene: Reaction kinetics and molecular weight distribution. In *J. Polym. Sci. A Polym. Chem.* 39, pp. 500– 513.

[65] Keii, T; Suzuki, E.; Tamura, M.; Murata, M.; Doi, Yoshihara (1982): Propene Polymerization with a Magnesium Chloride-Supported Ziegler Catalyst, 1. Principal kinetics. In *Makromol. Chem* 183 (10), pp. 2285–2304.

[66] Kahrman, R.; Erdogan, M.; Bilgic, T. (1996): Polymerization of propylene using a prepolymerized high-active Ziegler-Natta catalyst. I. Kinetic studies. In *J. Appl. Polym. Sci.* 60, pp. 333–342.

[67] Samson, J.J.C.; Weickert, G.; Heerze, A.E.; Westerterp, K. R. (1998): Liquid-phase polymerization of propylene with a highly active catalyst. . In *AlChE J.* 44 (6), pp. 1424–1437.

[68] Bonini, F.; Fraaije, V.; Fink, G. (1995): Propylene polymerization through supported metallocene/ MAO catalysts. Kinetic analysis and modelling. In *J. Polym. Sci. A Polym. Chem.* 33 (14), pp. 2393– 2402.

[69] Ochoteco, E.; Vecino, M.; Montes, M.; de la Cal, J. C. (2001): Kinetics and properties in metallocene catalysed propene polymerisations. In *Chem. Eng. Sci.* 56, pp. 4169–4179

[70] Alexiadis, A.; Andes, C.; Ferrari, D.; Korber, F.; Hauschild, K.; Bochmann, M.; Fink, G. (2004): Mathematical Modeling of Homopolymerization on Supported Metallocene Catalysts. In *Macromol. Mater. Eng.* 289 (5), pp. 457–466.

[71] Gonzalez-Ruiz, R. A.; Quevedo-Sanchez, B.; Laurence, R. L.; Henson, M. A.; Coughlin, E. B. (2006): Kinetic modeling of slurry propylene polymerization using rac-Et(Ind)$_2$ZrCl$_2$/MAO. In *AlChE J.* 52 (5), pp. 1824–1835.

[72] Huang, K.; Xie, R. (2014): Modeling of molecular weight distribution of propylene slurry phase polymerization on supported metallocene catalysts. In *Ind. Eng. Chem.* 20, pp. 338–344.

[73] Meier, G. B.; Weickert, G.; van Swaaij, Wim P. M. (2001): Comparison of Gas- and Liquid-Phase Polymerization of Propylene with Heterogeneous Metallocene Catalyst. In *J. Appl. Polym. Sci.* 81, pp. 1193–1206.

[74] Xie, T.; McAuley, K. B.; Hsu, J. C. C.; Bacon, D. W. (1994): Gas Phase Ethylene Polymerization. Production Processes, Polymer Properties, and Reactor Modeling. In *Ind. Eng. Chem. Res.* 33 (3), pp. 449–479.

[75] Shaffer, W. K. Al.; Ray, W. H. (1997): Polymerization of olefins through heterogeneous catalysis. XVIII. A kinetic explanation for unusual effects. In *J. Appl. Polym. Sci.* 65 (6), pp. 1053–1080.

[76] Soares, Joao B. P. (2001): Mathematical modelling of the microstructure of polyolefins made by coordination polymerization: a review. In *Chem. Eng. Sci.* 56, pp. 4131–4153.

[77] Touloupidis, V. (2014): Catalytic Olefin Polymerization Process Modeling: Multi-Scale Approach and Modeling Guidelines for Micro-Scale/Kinetic Modeling. In *Macromol. React. Eng.* 8 (7), pp. 508–527.

[78] Khare, N. P.; Lucas, B.; Seavey, K. C.; Liu, Y. A.; Sirohi, Ashuraj; Ramanathan, Sundaram et al. (2004): Steady-State and Dynamic Modeling of Gas-Phase Polypropylene Processes Using Stirred-Bed Reactors. In *Ind. Eng. Chem. Res.* 43 (4), pp. 884–900.

[79] McAuley, K. B.; MacGregor, J. F.; Hamielec, A. E. (1990): A kinetic model for industrial gas-phase ethylene copolymerization. In *AlChE J.* 36 (6), pp. 837–850.

[80] Kröner, T. (2014): Mass Transport and Kinetics in the Heterophasic Copolymerization of Propylene. Dissertation. Martin-Luther-Universität Halle-Wittenberg. Berlin: Mensch & Buch Verlag. ISBN 978-3-86387-527-5.

[81]  Kissin, Y. V.; Rishina, L. A. (2002): Hydrogen effects in propylene polymerization reactions with titanium- based Ziegler-Natta catalysts. I. Chemical mechanism of catalyst activation. In *J. Polym. Sci. A Polym. Chem.* 40, pp. 1353–1365.

[82]  Al-haj Ali, M.; Betlem, B.; Roffel, B.; Weickert, G. (2006): Hydrogen response in liquid propylene polymerization: Towards a generalized model. In *AIChE J.* 52 (5), pp. 1866–1876.

[83]  Samson, J. J. C.; Bosman, P. J.; Weickert, G.; Westerterp, K. R. (1999): Liquid-phase polymerization of propylene with a highly active Ziegler-Natta catalyst. Influence of hydrogen, cocatalyst, and electron donor on the reaction kinetics. In *J. Polym. Sci. A Polym. Chem.* 37, pp. 219–232.

[84]  Zakharov, V.A.; Bukatov, G.D.; Yermakov, Y.I. (1983): On the mechanism of olefin polymerization by Ziegler-Natta catalysts. In *Adv. Polym. Sci.* 51, pp. 61-100.

[85]  Begley, J. W. (1966): The role of diffusion in propylene polymerization. In *J. Polym. Sci. A-1 Polym. Chem.* 4 (2), pp. 319–336.

[86]  Schmeal, W. R.; Street, J. R. (1971): Polymerization in expanding catalyst particles. In *AIChE J.* 17 (5), pp. 1188–1197.

[87]  Singh, D.; Merrill, R. P. (1971): Molecular Weight Distribution of Polyethylene Produced by Ziegler-Natta Catalysts. In *Macromolecules* 4 (5), pp. 599–604.

[88]  Galvan, R.; Tirrell, M. (1986): Molecular weight distribution predictions for heterogeneous Ziegler-Natta polymerization using a two-site model. In *Chem. Eng. Sci.* 41 (9), pp. 2385–2393.

[89]  Hoel, E. L.; Cozewith, C.; Byrne, G. D. (1994): Effect of diffusion on heterogeneous ethylene propylene copolymerization. In *AIChE J.* 40 (10), pp. 1669–1684.

[90]  Yiagopoulos, A.; Yiannoulakis, H.; Dimos, V.; Kiparissides, C. (2001): Heat and mass transfer phenomena during the early growth of a catalyst particle in gas-phase olefin polymerization: the effect of prepolymerization temperature and time. In *Chem. Eng. Sci.* 56, pp. 3979–3995.

[91]  Bartke, M. (2002): Gasphasenpolymerisation von Butadien – Kinetische Untersuchungen im Minireaktor und Modellierung. Dissertation. Technische Universität Berlin. Berlin: Mensch & Buch Verlag. ISBN-10: 3898203344.

[92]  Patzlaff, M. (2006): Kinetische Untersuchung der Polymerisation von Propylen mit neuartigen Ziegler-Natta Katalysatoren in Gas- und Flüssigphase. Dissertation. Technische Universität Berlin.

[93]  Yermakov, Yu. I.; Mikhalchenko, V. G.; Beskov, V. S.; Grabovskii, Y. P. ; Emirova, I. V.  (1970): The role of transfer processes in gaseous phase polymerization of ethylene. In *Plast Massy*, 9, pp. 7-10 (original in Russian).

[94]  Crabtree, J. R.; Grimsby, F. N.; Nummelin, A. J.; Sketchley, J. M. (1973): The role of diffusion in the Ziegler polymerization of ethylene. In *J. Appl. Polym. Sci.* 17 (3), pp. 959–976.

[95]  Nagel, Eric J.; Kirillov, Valery A.; Ray, W. Harmon (1980): Prediction of Molecular Weight Distributions for High-Density Polyolefins. In *Ind. Eng. Chem. Prod. Res. Dev.* 19 (3), pp. 372–379.

[96]  Floyd, S.; Choi, K.-Y.; Taylor, T. W.; Ray, W. H. (1986): Polymerization of Olefins through Heterogeneous Catalysis. III. Polymer Particle Modelling with an Analysis of Intraparticle Heat and Mass Transfer Effects. In *J. Appl. Polym. Sci.* 32, pp. 2935–2960.

[97]  Floyd, S.; Choi, K.-Y.; Taylor, T. W.; Ray, W. H. (1986): Polymerization of Olefines through Heterogeneous Catalysis IV. Modeling of Heat and Mass Transfer Resistance in the Polymer Particle Boundary Layer. In *J. Appl. Polym. Sci.* 31 (7), pp. 2231–2265.

[98]  Hutchinson, R. A.; Ray, W. H. (1987): Polymerization of Olefins Through Heterogeneous Catalysis. VII. Particle Ignition and Extinction Phenomena. In *J. Appl. Polym. Sci.* 34, pp. 657–676.

[99]  Hutchinson, R.A.; Ray, W.H. (1990): Polymerization of olefins through heterogeneous catalysis. VIII. Monomer sorption effects. In *J. Appl. Polym. Sci.* 41, pp. 51–81.

[100] Hutchinson, R. A.; Chen, C. M.; Ray, W. H. (1992): Polymerization of olefins through heterogeneous catalysis X: Modeling of particle growth and morphology. In *J. Appl. Polym. Sci.* 44 (8), pp. 1389–1414.

[101] Guastalla, G.; Giannini, U. (1983): In *Makromol. Chem., Rapid Commun.* 4, pp. 519.

[102] Albizzati, E.; Galimberti, M.; Giannini, U.; Morini, G. (1991): In *Macromol. Chem., Macromol. Symp.* 48/49, pp. 223.

[103] Han-Adebekun, G. C.; Hamba, M.; Ray, W. H. (1997): Kinetic study of gas phase olefin polymerization with a TiCl4/MgCl2 catalyst I. Effect of polymerization conditions. In *J. Polym. Sci. A Polym. Chem.* 35 (10), pp. 2063–2074.

[104] Matsko, M.A.; Bukatov, G.D.; Mikenas, T.B.; Zakharov; V.A. (2001): *Macromol. Chem. Phys.* 202, pp. 1435–1439.

[105] Spitz, R.; Masson, P.; Bobichon, C.; Guyot, A. (1989): Activation of propene polymerization by hydrogen for improved MgCl2-supported Ziegler-Natta catalysts. In *Makromol. Chem.* 190 (4), pp. 717–723.

[106] Rishina, L.A.; Vizen, E.I.; Sosnovskaja, L.N.; Dyachkovsky, F.S. (1994): Study of the effect of hydrogen in propylene polymerization with the MgCl2-supported ziegler-natta catalyst—part 1. Kinetics of polymerization. In *Eur. Polym. J.* 30 (11), pp. 1309–1313.

[107] Mori, H.; Endo, M.; Tashino, K.; Terano, M. (1999): Study of activity enhancement by hydrogen in propylene polymerization using stopped-flow and conventional methods. In *J. Mol. Catal. A: Chem.* 145, pp. 153–158.

[108] Chadwick, J. C.; Miedema, A.; Sudmeijer, O. (1994): Hydrogen activation in propene polymerization with MgCl2-supported Ziegler-Natta catalysts: the effect of the external donor. In *Macromol. Chem. Phys.* 195, pp. 167–172.

[109] Blom, Richard; Dahl, Ivar M. (1999): On the sensitivity of metallocene catalysts toward molecular hydrogen during ethylene polymerization. In *Macromol. Chem. Phys.* 200 (2), pp. 442–449.

[110] Tsutsui, T.; Kashiwa, N.; Mizuno, A. (1990): Effect of hydrogen on propene polymerization with ethylenebis(1-indenyl)zirconium dichloride and methylalumoxane catalyst system. In *Makromol. Chem., Rapid Commun.* 11 (11), pp. 565–570.

[111] Busico, V.; Cipullo, R.; Corradini, P. (1993): Ziegler-Natta oligomerization of 1-alkenes: a catalyst's "fingerprint", 1. Hydrooligomerization of propene in the presence of a highly isospecific MgCl2-supported catalyst. In *Makromol. Chem.* 194 (4), pp. 1079–1093.

[112] Busico, V.; Cipullo, R.; Corradini, P. (1993): Ziegler-Natta oligomerization of 1-alkenes: A catalyst's "fingerprint", 2. Preliminary results of propene hydrooligomerization in the presence of the homogeneous isospecific catalyst system rac-(EBI)ZrCl2/MAO. In *Makromol. Chem., Rapid Commun.* 14 (2), pp. 97–103.

[113] Busico, V.; Cipullo, R.; Chadwick, J.C.; Modder, J. F.; Sudmeijer, O. (1994): Effects of Regiochemical and Stereochemical Errors on the Course of Isotactic Propene Polyinsertion Promoted by Homogeneous Ziegler-Natta Catalysts. In *Macromolecules* 27 (26), pp. 7538–7543.

[114] Vestberg, Torvald; Denifl, Peter; Parkinson, Matthew; Wilen, Carl-Eric (2010): Effects of external donors and hydrogen concentration on oligomer formation and chain end distribution in propylene polymerization with Ziegler-Natta catalysts. In *J. Polym. Sci. A Polym. Chem.* 48, pp. 351–358.

[115] Weickert, G.; Meier, G. B.; Pater, J. T. M.; Westerterp, K. R. (1999): The particle as microreactor: catalytic propylene polymerizations with supported metallocenes and Ziegler-Natta catalysts. In *Chem. Eng. Sci.* 54, pp. 3291–3296,

[116] Pater, J. T. M.; Weickert, G.; Loos, J.; van Swaaij, W. P. M. (2001): High precision prepolymerization of propylene at extremely low reaction rates—kinetics and morphology. In *Chem. Eng. Sci.* 56, pp. 4107–4120.

[117] Pater, J. T. M.; Weickert, G.; van Swaaij, W. P. M. (2003): Polymerization of liquid propylene with a fourth-generation Ziegler-Natta catalyst: Influence of temperature, hydrogen, monomer concentration, and prepolymerization method on powder morphology. In *J. Appl. Polym. Sci.* 87, pp. 1421–1435.

[118] Pater, J.T. M.; Weickert, G.; van Swaaij, W. P. M. (2003): Propene bulk polymerization kinetics: Role of prepolymerization and hydrogen. In *AIChE J.* 49 (1), pp. 180–193.

[119] Coutinho, F. M. B.; Costa, M. A. S.; Maria, L. C. S.; Bruno, J. C. (1994): Particle control of Ziegler–Natta catalysts based on TiCl3 for propylene polymerization. Effect of prepolymerization. In *J. Appl. Polym. Sci.* 51 (6), pp. 1029–1034.

[120] Soares, Joao B.P.; Hamielec, Archie E. (1996): Effect of hydrogen and of catalyst prepolymerization with propylene on the polymerization kinetics of ethylene with a non-supported heterogeneous Ziegler-Natta catalyst. In *Polymer* 37 (20), pp. 4599–4605.

[121] Chu, K.-J.; Soares, J. B.P.; Penlidis, A.; Ihm, S.-K. (2000): Effect of prepolymerization and hydrogen pressure on the microstructure of ethylene/1-hexene copolymers made with MgCl2-supported TiCl3 catalysts. In *European Polymer Journal* 36 (1), pp. 3–11.

[122] Monji, M.; Abedi, S.; Pourmahdian, S.; Taromi, F. A. (2009): Effect of Prepolymerization on Propylene Polymerization. In *J. Appl. Polym. Sci.* 112 (4), pp. 1863–1867.

[123] Zacca, J. J.; Debling, J. A. (2001): Particle population overheating phenomena in olefin polymerization reactors. In *Chem. Eng. Sci.* 56 (13), pp. 4029–4042.

[124] Kröner, S.; Eloranta, K.; Bergstra, M. F.; Bartke, Michael (2007): Kinetic Study of the Copolymerisation of Ethylene with a Single Site Catalyst in Propane Slurry Polymerisation. In *Macromol. Symp.* 259, pp. 284–294.

[125] Piduhn, M.(1999): Metallocen-katalysierte Polymerisation von Propen und Ethen aus der Gasphase. Berlin: Mensch-und-Buch-Verlag. ISBN: 978-3898200240.

[126] Al-haj Ali, M.; Betlem, B.; Roffel, B.; Weickert, G. (2007): Estimation of the Polymerization Rate of Liquid Propylene Using Adiabatic Reaction Calorimetry and Reaction Dilatometry. In *Macromol. React. Eng.* 1 (3), pp. 353–363.

[127] Zogg, A.; Stoessel, F.; Fischer, U.; Hungerbühler, K. (2004): Isothermal reaction calorimetry as a tool for kinetic analysis. In *Thermochim. Acta* 419, pp. 1–17.

[128] Korber, F.; Hauschild, K.; Fink, G. (2001): Reaction Calorimetric Approach to the Kinetic Investigation of the Propylene Bulk Phase Polymerization. In *Macromol. Chem. Phys.* 202 (17), pp. 3329–3333.

[129] Korber, F.; Hauschild, K.; Winter, M.; Fink, G. (2001): Reaction Calorimetric Investigation of the Propylene Slurry Phase Polymerization with a Silica-Supported Metallocene/MAO Catalyst. In *Macromol. Chem. Phys.* 202 (17), pp. 3323–3328

[130] ChemiSens AB (2007): ChemiSens Reaction Calorimetric Systems. The Chemical Process Analyser CPA202. ChemiSens document Z202-002. Edited by ChemiSens AB, Lund, Sweden.

[131] Marx, R. (2008): Realization of Reaction Setup for Olefin Polymerizations. Diploma Thesis. TU Dresden. Dresden.

[132] National Institute of Standards and Technology (2016): NIST Chemistry WebBook. NIST Standard Reference Database Number 69. National Institute of Standards and Technology. USA. Available online at http://webbook.nist.gov/chemistry/.

[133] ASTM D1238-13. Standard Test Method for Melt Flow Rates of Thermoplastics by Extrusion Plastometer. ASTM International. West Conshohocken, PA. 2013.

[134] ISO 1133-1:2011. Plastics -- Determination of the melt mass-flow rate (MFR) and melt volume-flow rate (MVR) of thermoplastics -- Part 1: Standard method. International Organization for Standardization. Switzerland. 2011.

[135] Blaine, R. L.: Thermal applications note. Polymer heats of fusion. TA Instruments. New Castle DE, USA. Available online at http://www.tainstruments.com/pdf/literature/TN048.pdf.

[136] Verein Deutscher Ingenieure, VDI-Gesellschaft Verfahrenstechnik und Chemieingenieurwesen (GVC) (2002): VDI-Wärmeatlas. Berechnungsblätter für den Wärmeübergang. Neunte, überarbeitete und erweiterte Auflage. Berlin, Heidelberg: Springer Berlin Heidelberg. ISBN: 978-3-662-10744-7.

[137] Brandrup, J.; Immergut, E. H.; Grulke, E. A. (2005): Polymer handbook. 4. ed. Norwich, NY: Knovel (A Wiley-interscience publication). ISBN: 0471166286.

[138] Hutchinson, R. A.; Ray, W. H. (1991): Polymerization of olefins through heterogeneous catalysis - the effect of condensation cooling on particle ignition. In *J. Appl. Polym. Sci.* 43 (7), pp. 1387–1390.

[139] Ranz, W. E.; Marshall, W. R. (1952): Evaporation from drops. I. In *Chem. Eng. Prog.* 48, pp. 141–146.

[140] Bartke, M.; Reichert, K.-H. (1999): Berechnung von Molmassenverteilungen bei Polyreaktionen mit Standard-Simulationsprogrammen. In *Chem. Ing. Tech.* 71, pp. 1310–1314.

[141] Stern, S. A.; Mullhaupt, J. T.; Gareis, P. J. (1969): The effect of pressure on the permeation of gases and vapors through polyethylene. Usefulness of the corresponding states principle. In *AIChE J.* 15 (1), pp. 64–73.

[142] Process Systems Enterprise Ltd. (2015): gProms ModelBuilder Documentation. Release 4.2.0. Edited by Process Systems Enterprise Ltd. London, UK. Available online at https://www.psenterprise.com/products/gproms/modelbuilder.

[143] Choi, Kyu-Yong; Ray, W. Harmon (1985): Recent Developments in Transition Metal Catalyzed Olefin Polymerization–A Survey. II. Propylene Polymerization. In *J. Macromol. Sci., Rev. Macromol. Chem. Phys.* 25 (1), pp. 57–97.

[144] ChemiSens AB (2011): User manual CPA202 system hardware. Edited by ChemiSens AB. Lund, Sweden.

[145] Flory, P. J. (1953): Principles of Polymer Chemistry. Ithaka, New York: Cornell University Press. ISBN: 0801401348.

[146] Barton, A. F. M. (1990): CRC handbook of polymer-liquid interaction parameters and solubility parameters. Boca Raton, Florida: CRC Press. ISBN: 0849335442.

[147] Bradford, M. L.; Thodos, G. (1966): Solubility parameters of hydrocarbons. In *Can. J. Chem. Eng.* 44 (6), pp. 345–348.

[148] Aminabhavi, T. M.; Phayde, H. T.S.; Ortego, J. D. (1996): A Study of Sorption/Desorption and Diffusion of n-Alkanes and Aliphatic Hydrocarbons into Polymeric Blends of Ethylene-Propylene Random Co-Polymer and Isotactic Polypropylene in the Temperature Interval of 25 to 70°C. In *J. Polym. Eng.* 16 (1-2), pp. 121–148.

[149] Williams, R. B.; Katz, D. L. (1954): Vapor Liquid Equilibria in Binary Systems. Hydrogen with Ethylene, Ethane, Propylene, and Propane. In *Ind. Eng. Chem.* 46 (12), pp. 2512–2520.

[150] Mizan, T. I.; Li, J.; Morsi, B. I.; Chang, M.-Y.; Maier, E.; Singh, C. P. P. (1994): Solubilities and mass transfer coefficients of gases in liquid propylene in a surface-aeration agitated reactor. In *Chem. Eng. Sci.* 49 (6), pp. 821–830.

# Curriculum Vitae

## Personal data

Name:                          Dipl.-Ing. Joana Kettner

Date and place of birth:       12.09.1985 in Bautzen, Germany

Gender:                        Female

Citizenship:                   German

## Education

11/2011 – 03/2016    **Ph.D. study in Polymer Reaction Engineering**

Martin-Luther-University Halle-Wittenberg, Faculty of Natural Sciences II - Chemistry, Physics and Mathematics, Institute of Chemistry, Workgroup of Prof. Dr.-Ing. Bartke - Polymer Reaction Engineering

Subject: Kinetic investigation of different supported catalysts in polymerization of propylene under industrially relevant conditions

10/2004 – 02/2010    **Diploma study Process Engineering**

Technical University Bergakademie Freiberg, Faculty of Mechanical, Process and Energy Engineering

Field of study: Chemical and biological process engineering

Diploma thesis: "Homogene Methanoxidation in keramischen Schwämmen" (Homogeneous methane oxidation in ceramic sponges)

11.02.2010           **Diplom-Ingenieur Process Engineering**

09/1996 – 07/2004    **General qualification for university entrance**

Schillergymnasium Bautzen

03.07.2004           Certificate: Abitur

## Professional experience

Since 09/2016        **Lead Scientist** at Borealis Polyolefine GmbH, Linz

Research and development in the field of olefin polymerizations and catalysts, high throughput experimentations